£20.

Animal models for psychiatry

J.D. Keehn

Professor of Psychology, Atkinson College
York University, Toronto

Animal models for psychiatry

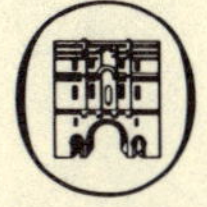

ROUTLEDGE & KEGAN PAUL
London, Boston and Henley

First published in 1986
by Routledge & Kegan Paul plc

14 Leicester Square, London WC2H 7PH, England

9 Park Street, Boston, Mass. 02108, USA and

Broadway House, Newtown Road,
Henley on Thames, Oxon RG9 1EN, England

Set in 10 on 11pt Baskerville
by Columns of Reading
and printed in Great Britain
by Billing and Sons Ltd.
Worcester.

Library of Congress Cataloging in Publication Data
Keehn, J. D.
Animal models for psychiatry.
Bibliography: p.
Includes index.
1. Mental illness — Animal models. 2. Psychiatry —
Research — Moral and ethical aspects. I. Title.
RC337.K44 1985 616.89'00724 85-2174

British Library CIP data also available

ISBN 0-7102-0562-7

Contents

Figures and Tables

Figures

Tables

Preface

Psychiatry today is in a state of psychosis. On the one hand there is depression and despair over the status of diagnosis and the actuality of mental illness, while on the other hand there is elation over biochemical advances in the understanding of schizophrenia; there is optimism about the pharmacological control of psychotic syndromes and at the same time pessimism about the long-term side effects of these drug therapies; and, finally, there is relief at the success of medication in the battle with anxiety, combined with anxiety over signs that the medication is addicting. Meanwhile psychology is in no less a psychotic condition, being unable to make up its mind whether or not such things as minds exist to be made up. Out of these predicaments a new psychiatric speciality, biological psychiatry has blossomed.

Biological psychiatry is not the study of the phenomenology of human mental illness, but draws its substance from the biological sciences, behavioural pharmacology and the sciences of animal behaviour, especially. With its discoveries it provides the clinical psychiatrist and psychologist with means of analysing and treating the sources and manifestations of human miseries. This book describes some of these miseries and some of the work in hand to relieve them. The focus is on animal contributions to the study of human 'psychopathology', but examples of natural animal behavioural abnormalities are also given. It is important to understand that certain afflictions of humans are also afflictions of lower animals, and that insofar as we share a common biological origin with these animals, then whatever is learned about the one necessarily adds to our knowledge about the other.

Chapter 1 includes some behavioural disorders in animals, of a non-experimental origin, that resemble human disorders under similar circumstances. With the animals the particular circumstances are the immediate focus of interest, but with humans a personality malfunction is more often sought. The first chapter also includes an account of some ways in which such malfunctions have been considered, particularly from the standpoint of mental illnesses and clinical pictures. Criteria and pitfalls with experimental models of these conditions are also considered in this chapter.

Chapter 2 takes up the questions of animal suffering and ethics of animal experimentation. Methods of humane experimentation with animals are described, along with an account of animal and human natures. Chapters 3 and 4 survey various behavioural anomalies found in laboratory animals and relate them to comparable human disablements. The next four chapters examine animal contributions to the analysis of human neurosis and psychophysiological disorders, addictions, psychosis and disorders of childhood.

In none of these categories are complete 'hi-fidelity' models of human psychiatric conditions looked for. The relevance and value of animal studies varies from case to case: with psychosis it is to do with drugs and hospital management; with neurosis, with learning; with childhood disorders, with maternal deprivation and development; and with addictions, the roles of conditioned withdrawal and tolerance in relapse after detoxification.

Thus the present work is not an attempt to force animal models on to clinical psychiatry but to illustrate ways in which these models can have psychiatric value. A summary of such ways constitutes the final, concluding chapter.

At this point I would like to acknowledge the secretarial assistance provided by Jackie Logan, Rischa Sidon, Rhonda Strasberg and Marilyn Weinper, whose contributions are distributed throughout the entire manuscript, and to thank those authors and publishers who have permitted me to reproduce figures and tables from their original works. Appropriate acknowledgments in these cases appear with each individual illustration.

J.D.K. Toronto, 1985

Part I
Domains of biological psychiatry

1 Psychopathology: the status of animals

Comparative experimental psychiatry

Scope and origin

A few years ago, Gay (1967) made a vigorous plea for expanding the domain of comparative medicine, and two years later Leader (1969) surveyed the long history of contributions to medical skills made by comparative pathologists seeking animal models of human diseases. Soon after that, three famous ethologists, Karl von Frisch, Konrad Lorenz and Nikolas Tinbergen received a Nobel prize for the lessons that their accounts of animal habits have taught medical practitioners about human diseases and behaviour disorders. In his acceptance speech, Tinbergen (1974) vividly recounts how the ethological methods of studying natural behaviours of animals enlarged his understanding of the behaviour of autistic children and contributed to new methods of treating this childhood disability.

The psychological study of abnormal behaviour in animals does not have as long and impressive a history as that of the study of physical diseases by way of laboratory animal models, nor has it attained the eminence accorded to ethological research; but it has amassed a substantial body of information since its beginning with the work of Maria Yerofeeva (Erofeew, 1916) in Pavlov's laboratory in 1912. Following the pioneering work of Pavlov (1927) and his colleagues there have been numerous attempts to generate animal models of human behavioural anomalies. These include early studies of cats by Masserman (1943) and Wolpe (1958), of dogs by Gantt (1944) and of sheep and goats by Liddell (1944), all of which were directed to the creation of animal versions of human neuroses. It is now doubtful that these early efforts achieved their intentions (Broadhurst, 1973), but the number of human psychological disabilities for which animal models have since been proposed include schizophrenia (Ellison, 1979), depression (Seligman, 1975), asthma (Ottenberg, Stein, Lewis and Hamilton, 1958), epilepsy (Gaito, 1976; Killam, Killam and Naquet, 1967), infantile autism (Scott and Senay, 1973), alcoholism and drug addiction (Gilbert, 1977), and many more (Keehn, 1979; Maser and Seligman, 1977).

Some of these conditions are the results of deliberate attempts to create specific parallels to human problems, while others emerged unexpectedly from routine care and maintenance laboratory practices (Startsev, 1976), or developed in the course of investigations into basic behavioural processes. The classic example of experimental neurosis first appeared in this way, and a similar later case is recounted by Ellison (1979):

> I became interested in . . . experimental neurosis . . . when . . . one of the dogs I was training in a difficult discrimination developed a classical case. This dog had merely to stand in a conditioning platform and would receive a piece of dog candy 16 sec after brief presentation of a high tone but no candy following a brief low tone. He gradually refused to come to the experiment and had to be carried, even though dogs being shocked will readily follow the experimenter. He also developed trembling attacks and refused to eat candy dispensed from the feeder although he was quite hungry and would eat from the experimenter's hand (unless the piece of candy had previously been dispensed from the feeder!). His salivation became quite erratic, and he would stare fixedly at spots on the wall. (Ellison, 1979, p.82)

It may be improper to call this animal neurotic in the human sense (Hunt, 1964), or even to describe either animals or humans as psychopathological. Psychopathology, however, is a convenient shorthand for alluding to behavioural abnormalities that have been recognized over the ages, and is used here for contemporary convenience, like psychology and psychosomatics, with only token commitment to a medical model of madness (Siegler and Osmond, 1974).

What must an animal model?

Disease entities, diagnosis and clinical pictures

A major function of psychiatry is the diagnosis of mental illnesses associated with particular patterns of abnormal behaviours, perceptions and emotions, although voices have been raised against such undertakings. Thus, according to Frank (1975):

> What we truly need, in the study of psychopathology, is a revolution. . . . Without it, we . . .are doomed to an outmoded . . . inappropriate mode of conceptualization that contributed little if anything to the understanding of people. Our system of classification in psychiatry has proven to be: full of sound and fury, signifying nothing. (Frank, 1975, p.80)

But others, like Kendell (1975), express another opinion.

> Many present day psychiatrists have lost interest in the whole issue of diagnosis while others have suggested that it is an unnecessary, even a harmful exercise. This book was born of the conviction that such attitudes are profoundly mistaken, and that the development of a reliable and valid classification of the phenomena of mental illness are two of the most important problems facing contemporary psychiatry. (Kendell, 1975, p. vii)

These problems have been faced before (Menninger *et al.*, 1963). Several solutions have been offered, including one by Carl Wernicke (1848–1905) who distinguished deficit, distortion and excess in sensory, motor and cognitive functions according to the following scheme:

	Function		
	Psychosensory	*Psychomotor*	*Intrapsychic*
Deficit	Anaesthesia	Akinesia	Afunction
Excess	Hyperaesthesia	Hyperkinesia	Hyperfunction
Distortion	Parasthesia	Parakinesia	Parafunction

This schema permits a nine-fold classification of possible psychological malfunctions, but it is not a diagnostic system for differential diagnosis of mental diseases in the modern fashion. Numerous other classifications have been proposed throughout the ages. These are summarized by Menninger *et al.* (1963) who placed them in two groups that derive, respectively, from the historic traditions of Hippocrates and Plato:

1 classifications of individuals into clinical types based on presenting clinical pictures; and
2 classifications of disease entities.

In the Hippocratic tradition, the emphasis is on case histories of individuals and on treatment formulations based on similarities in clinical pictures of different patients; in the Platonic tradition, the emphasis is on categorical diagnosis, based on the assertion that disease entities are ideal universals of which particular patients are more or less imperfect examples. Modern institutional psychiatry is mainly Platonic, while psychoanalysis, behavioural analysis and allied movements are Hippocratic.

The two major classes of modern arrangements of mental disorders in the Platonic sense, neuroses and psychoses, began to take their forms in the seventeenth and eighteenth centuries. That part of a botanical system to do with neurosis devised by William Cullen (1712–90) is summarized in Table 1.1.

Table 1.1 William Cullen's classification of neuroses

Class:	*Neurosis* –	disturbances of sense and motion without infection.
Orders:	I	*Diminution of voluntary motion or sense* with 2 genera: apoplexy and paralysis.
	II	*Diminution of involuntary motions* with 4 genera, including vomiting and dyspepsia.
	III	*Irregular motions of muscles and spasms* with 17 genera, including epilepsy, asthma, chorea and hysteria.
	IV	*Disorders of judgment without fever or coma* with 4 genera, including amentia, mania and melancholia

Adapted from Menninger, K., *et al.* (1963) *The Vital Balance*, New York, Viking.

In the same era Pinel (1745–1826) proposed a classification by clinical types that now would be attributed to psychosis or organic brain disorder.

Pinel's fundamental clinical types

1 Mania: acute excitement.
2 Melancholia: depression.
3 Dementia: incoherent thought.
4 Idiotism: feeble-mindedness.

The basis of present classification systems was laid down by Kraepelin (1856–1926), and after several attempts to accommodate psychiatric disorders with international classifications of diseases and causes of death, the American Psychiatric Association (1952) proposed a complex diagnostic system based on psychiatric experience in World War II. A simplification of the Association's

Table 1.2 Classification of abnormal patterns of behaviour

Disorders caused by or associated with impairment of brain tissue function

A Acute brain disorders	C Mental deficiency
B Chronic brain disorders	1 Familial or hereditary deficiency
	2 Idiopathic deficiency

Disorders of psychogenic origin or without clearly defined physical cause or structural change in the brain

Psychotic disorders	*Neurotic disorders*
A Involutional psychosis	A Anxiety reaction
B Affective reactions	B Dissociative reaction
C Schizophrenic reactions	C Conversion reaction
1 Simple	D Phobic reaction
2 Hebephrenic	E Obsessive-compulsive reaction
3 Catatonic	F Depressive reaction
4 Paranoid	G Other neurotic reactions
5 Acute undifferentiated	
6 Chronic undifferentiated	
7 Schizo-affective	
8 Childhood	*Character disorders*
D Paranoid reactions	A Personality pattern disturbance
	B Personality trait disturbance
Psychosomatic disorders	C Sociopathic personality
A Skin reaction	D Special symptoms
B Musculoskeletal reaction	
C Respiratory reaction	
D Cardiovascular reaction	*Transient personality disorders*
E Haemic and lymphatic reaction	A Gross stress reaction
F Gastrointestinal reaction	B Adult situation reaction
G Genitourinary reaction	C Adjustment reaction of infancy
H Endocrine reaction	D Adjustment reaction of childhood
I Nervous system reaction	E Adjustment reaction of adolescence
J Special sense reaction	F Adjustment reaction of late life

Simplification of a classification of psychological abnormalities proposed by the American Psychiatric Association based upon experience during and after World War II. (*Diagnostic and Statistical manual, Mental Disorders*, Washington, American Psychiatric Association, 1952.) Unlike earlier classifications made by individual men, this classification was agreed upon by a committee of psychiatrists. The classification was expanded in *DSM-II* and broadened into a multi-axial system in *DSM-III*. See text for details (Reproduced from Keehn, J.D. (1962) *The Prediction and Control of Behavior: A Shorter Introduction to Psychology*, Beirut, Khayat).

Table 1.3 Multi-axis diagnostic and statistic system adopted by the American Psychiatric Association (DSM-III)

AXIS I	*Traditional psychiatric syndrome* Principally administrative diagnosis and allocation for treatment, epidemiology and statistics.
AXIS II	*Personality disorders (adults); Developmental disorders (children)* For comparisons with norms rather than with ideals.
AXIS III	*Non-mental medical disorders* Usual physical medical history.
AXIS IV	*Severity of stress in previous year* Rated on a 7-point scale.
AXIS V	*Highest level of adaptive behavior in previous year* Rated on a 7-point scale.

This multi-axis system has replaced the 'botanical' classification system employed in *DSM-I* (see summary in Table 1.2) and its successor *DSM-II*. Already a committee is considering a revision that will become *DSM-IV*. A brief psychological appraisal of *DSM-III* appears in Schact, T., and Nathan, P.E. (1977) 'But is it good for psychologists? Appraisal and status of DSM-III', *American Psychologist*, *32*, 1017–25. The system is described in full in Webb, L.J., *et al.* (1981) DSM-III training guide, New York, Brunner Mazel.

official classification (*Diagnostic and Statistical Manual I*) is shown in Table 1.2. Since that classification the World Health Organization has revised its International Classification of Diseases, and the American Psychiatric Association's system has passed through a second (*DSM-II*) to a third (*DSM-III*) revision. The third revision combines diagnosis and formulation by including the categorization of diseases and the characterization of persons according to the summary in Table 1.3. From it, treatment plans are based on personal case histories (in the Hippocratic tradition) rather than on diagnostic labels that mark universal Platonic-type disease entities. The third diagnostic and statistical manual of the American Psychiatric Association thus incorporates both historical aspects of diagnosis in mental illness. Axis I represents the traditional psychiatric diagnostic system, although revised and expanded to include disorders of social communication as well as the mental illnesses summarized in Table 1.2. This axis is relevant for animal models but may turn out to be less important than Axis II, which is to do with atypical childhood and adult behaviours.

As an alternative to the formal categorization of mental illnesses, Shapiro (1975) proposes a list of ten manifestations of human psychopathology marked by a combination of distress, disablement, incongruence with reality and social inappropriateness.

1 *Intense referential feelings*, such as fear of dogs and depression at work.
2 *Intense non-referential feelings*, like generalized anxiety and depression.
3 *Exaggerated or reduced drives*, expressed as insomnia, impotence and obesity.
4 *Strong irrational beliefs*, such as jealousy and paranoia.
5 *Cognitive dysfunctions*, like inability to concentrate or remember.
6 *Maladapted non-social behaviours*, in such forms as persistent handwashing and housecleaning.
7 *Maladapted social behaviours*, characterised by abusive language, talking to oneself and withdrawal.
8 *Disturbed perceptual experience*, such as hallucinations.
9 *Very intense somatic experiences*, like intense headaches, asthma and gastro-intestinal pains.
10 *Motor dysfunctions*, including tics, tremors, dyskinesias and stereotypies.

Plainly some of these manifestations of human psychopathology are closer to the psychology of animals than are others. Motor disturbances are easily observed, and very intense somatic disturbances may be readily inferred in animals, as may be reduced or exaggerated drives, but with intense referential feelings, disturbances of perception and cognition and strong irrational beliefs it is a very different matter, with maladaptive social and non-social behaviours somewhere between the extremes. Nevertheless, illustrations of some of these characteristic human psychological problems are available in animals. A case of bizarre posturing and possible hallucination by an isolation-reared monkey is reproduced in Chapter 3, where unnatural animal movements and postures are described in detail. In addition, phobias and fears are often attributed to laboratory animals that succeed in avoiding electric shocks in standard laboratory situations. Animals that fail to avoid, on the other hand, after prior exposure to inescapable shock, are said by Seligman (1975) to fail through a learned sense of helplessness. The fear and the helplessness supposedly felt by the animals in these cases may or may not experientially resemble those that Shapiro (1975) identifies as common disabling and distressing features in humans, but these are matters of symptomatology, in which humans and lower animals cannot be expected to be identical. Nevertheless, originating circumstances that lead to behavioural disturbances in animals may be similar to those that cause psychopathologies in people.

Symptomatology and origination

Uncertainty about the function of diagnosis in psychiatry greatly complicates the search for animal models of human mental illness by failing to provide clear objects for the models to copy, but Broadhurst (1973) may have reached an unnecessarily harsh conclusion by taking one side of the disease entity *versus* clinical picture debate.

> In 1960 it seemed sufficient to explore the extent of the evidence for an experimental neurosis in the strict sense, that is to say, to decide if evidence for a clear-cut animal analogue of a clinical disease entity of some sort existed. The answer, not surprisingly, it now seems was 'No'. No animal preparation that convincingly mimicked such entity could be provided by experimental psychology for the study of etiology, prognosis or therapy. (Broadhurst, 1973, p. 745)

As it happens, many of the early attempts to generate experimental neurosis in laboratories actually demonstrated animal clinical pictures. Audiogenic seizures in mice and rats (Krushinskii, 1962), tonic immobility in chickens, rabbits, lizards, toads and goats (Crawford and Prestrude, 1977), and fear-induced feeding problems in laboratory cats (Masserman, 1943; Wolpe, 1958) and dogs (Pavlov, 1927) are prominent examples.

Evidence for an experimental neurosis may be non-existent for the reason that there is no natural object for an experimental image to depict. Recent studies of experimental neurosis from Russian origins (Miminoshvili, 1960; Startsev, 1976) focus more on specific psychophysiological disorders like gastric achylia, hyperglycaemia, sexual dysfunction, hypertension and cardiac insufficiency than on general neurotic emotionality, and this focus is generally true whenever modern animal analogues of human disabilities are sought. Thus not schizophrenia but the biochemical roots of stereotypy is the basis of some animal models of psychosis; not alcoholism but relationships between animal and human responses to alcohol; not anaclitic depression but the animal and human reactions to isolation rearing, are the foci of contemporary research on animal models in psychopathology.

Even so, as with psychiatric indecision about whether to keep or discard traditional diagnosis, laboratory workers with animals still cautiously continue using traditional terms. Classification systems as elaborate as those of the American Psychiatric Association and World Health Organization are too complex ever to be of use with animals, but a more manageable list of diagnostic categories is employed by Woodruff, Goodwin and Guze (1974). All of the twelve psychiatric diagnoses that they enumerate and describe – affective disorders, schizophrenic disorders, anxiety neurosis, hysteria, obsessional neurosis, phobic neurosis, alcoholism, drug dependence, sociopathy, brain syndrome, anorexia nervosa and sexual problems – have counterparts in animal behaviours. As with humans, there are reports of animal affective disorders (Seligman, 1975), schizophrenic disorders (Ellison, 1979), anxiety neurosis (Pavlov, 1927), hysteria (Sanger and Hamdy, 1962), obsessional neurosis (Ellen, 1956), phobic neurosis (Wolpe, 1958), alcoholism (Falk, Samson and Winger, 1972), drug dependence (Jones and Prada, 1973), sociopathy (Ellison, 1979), brain syndrome (Auer and Smith, 1940), anorexia nervosa (Masserman, 1943) and animal problems with sex (Chertok and Fontaine, 1963). Nevertheless, comparative experimental psychiatry cannot progress on the basis of presenting symptoms alone, for as Liddell (1956) long ago remarked, a psychodynamics of animal behaviour can never be. He continues:

> Because of man's incredibly complicated cognitive machinery, his neurotic symptoms may exhibit a bewildering diversity. Nevertheless, our emotionally disturbed animals under careful observation show many of the same or closely similar symptoms. The physician is in a much better position, however, to explore and analyze his psychoneurotic patient's symptomatology than is the behaviorist in the case of his experimentally neurotic animal. When it comes to analyzing the *originating situations* . . . the shoe is on the other foot. The behaviorist can *create* and rigorously control the situation in which experimental neuroses originate. (Liddell, 1956, p. 59)

Liddell spoke only of experimental neurosis, but the point applies to any animal model of human psychopathology. In this and other chapters many spontaneous animal disablements are described with psychiatric labels, but the human and animal cases can never be exactly the same in symptomatology. However, research scientists in this field run not one risk but two: one that the disablements they create are specific to their laboratory subjects, non-existent in natural animal life; the other that their creations are inapplicable to human disorders. Let me address the first risk by way of examples of spontaneous abnormal animal behaviour, and the second by way of criteria for animal models of human psychopathologies.

The psychopathology of animal life

Spontaneous animal psychopathologies

In farms and zoos

Abnormal animal behaviours that correspond to psychopathologies in humans are not only found in laboratories as deliberate creations for the benefit of mankind. Behavioural abnormalities spontaneously exhibited by all sorts of animals have been observed by ethologists, zoologists, veterinarians, husbandrymen, farmers and animal keepers in circuses and zoos (Fox, 1968). Thus there are reports of intromission phobia in the bull (Fraser, 1957), hysteria in hens (Sanger and Hamdy, 1962), ulcerative colitis in the gibbon (Stout and Snyder, 1969), and sexual inversions in certain birds and fish (Morris, 1955).

Chertok and Fontaine (1963), in an account of psychosomatics in veterinary medicine, include collective epilepsy among dogs, nymphomania in cats, post-emotional traumas in horses, dogs and cats, and impotence and pseudopregnancy in the rat. To these, Brion (1964) and Levy (1952) add a variety of animal tics, convulsions and seizures, while Fraser (1960) describes cases of spontaneous tonic immobility ('animal hypnosis') in the horse, the goat and the cow. Several more behavioural problems of farm animals, described in a standard textbook of veterinary medicine as 'functional nervous diseases in animals', are summarized in Table 1.4.

Behavioural abnormalities of captive animals in zoos include non-adaptive escape reactions, food refusal, excessive aggression, stereotyped motor reactions, displacement scratching, self-mutilation, homosexuality, sexual perversions, perversions of appetite (coprophagia), apathy, and defective mother–infant interactions (Meyer-Holzapfel, 1968). Among other possibilities, Meyer-Holzapfel interprets these as results of interference with normal inter-animal social intercourse, and of thwarting of natural flight reactions when danger signals appear. These could explain the sudden deaths of zoo animals described below, and also the arrested development of Merlin, the wild orphaned chimpanzee that van-Lawick Goodall (1971) observed.

In the wild

Merlin was a 3-year-old chimpanzee who was 'adopted' by his older sister Miff after his mother died. He was involved in a violent

Table 1.4 Functional nervous diseases in animals

1 Dizziness (vertigo) in horses and dogs.
2 Epilepsy (falling sickness) in dogs.
3 Tetany (muscular spasms) in various animals.
4 Catalepsy, e.g. convulsions, muscular plasticity in cows.
5 Neuroses of pregnancy, parturition and lactation; milk fever (coma).
6 Chorea, complex involuntary arrhythmic muscular movements.
7 Twitch spasms (tics).
8 Muscular tremors.
9 Psychoses: feeblemindedness
traumatic dementia
maniacal staggers in horses
mass panic in herds (stampeding)
hysteria in dogs (fright disease, running fits)
degenerative psychopathic constitution (sexual perversions)

Condensed from Hutyra, F., Marek, J., and Manninger, R. (1949) *Special Pathology and Therapeutics of the Diseases of Domestic Animals*, vol. III, 5th edition, London, Balliere, Tindall & Cox.

encounter with a dominant male whose charge he failed to avoid, and thereafter developed abnormal social behaviour and a bizarre autistic stereotypy.

MERLIN

> When he was four years old Merlin was far more submissive than other youngsters of that age: constantly he approached adults to ingratiate himself, turning repeatedly to present his rump, or crouching, pant-grunting before them. At the other end of the scale, Merlin was extra-aggressive to other infants of his own age. . . .
>
> As Merlin entered his sixth year his behavior was becoming rapidly more abnormal. Sometimes he hung upside down . . . suspended almost motionless for several minutes at a time. Hunched up with his arms around his knees, he sat often rocking from side to side with wide-open eyes. (van Lawick-Goodall, 1971, p. 227)

Illustrations such as Merlin are uncommon because naturalistic observations of wild animals more often focus on herds or troops than on target individuals, and abnormality is normally an individual, not a collective, phenomenon. Nevertheless collective behavioural pathologies have been seen in humans, e.g. St Vitus's dance and mob hysteria, and there are comparable situations with animals. Migrating lemmings and stampeding cattle are well-known examples; less well known are the rampages of herds of elephants intoxicated through consumption of fermented grain and fruit (Carrington, 1959).

During care and maintenance routines

As with Merlin, disturbances in social relationships can cause pathological responses by caged animals in laboratories and zoos, even when they are not exposed to deliberate stress. The cases of agitation in the rhesus monkey, Cupid, and the chimpanzee, Dennis, described in Chapters 2 and 7, respectively, are illustrations. Both were separated from their normal mates for routine laboratory purposes. Similar examples from zoos, leading to inexplicable sudden deaths, are reported by Christian and Ratcliffe (1952) for otters, minks, a cheetah, a serval and a lynx, and by Stout and Snyder (1969) for Saimang gibbons. From the Sukhumi Research Station in the USSR, Startsev (1976) reports the sudden death of a baboon following a routine cage reshuffle that brought the victim in view of a group of dominant males.

ZAGREB

> On August 7, 1965 [Zagreb and Ambarchik] were simultaneously transferred to a large cage. . . . Outside the cage, separated by a transparent barrier, were several full grown males, who greeted their new young neighbours with threatening gestures and vocalizations. . . . In the very first minutes both showed a gait disturbance characterized by incoordination and a posture with the knees half flexed. . . . During the subsequent 3 days [Zagreb's] movements reflected a steadily progressive hypotonia: he sat continuously with arms extended and head sunk down between his knees. . . . [Later] one of the animal handlers attempted to catch him . . . and he fell to the floor, struggled a short time, and died. A careful autopsy revealed no anatomical lesion to which one could attribute such severe motor impairment. (Startsev, 1976, pp.131–2)

Ewing (1967) reports a similar phenomenon in cockroaches, wherein the subordinate members of pairs may die with no signs of physical injury when they are repeatedly attacked by dominant partners who do not respond to the victim's submissive posture.

Table 1.5 Most common spontaneous neuropathological disorders observed in a group of thirty-nine hamadryas baboons held in captivity

Disorder	*Number of animals*
Tonic-clonic seizures	15
Adynamia after and between attacks	14
Vomiting	14
Tonic seizures	13
Fibrillary muscle twitches	11
Tremors	10
Myorhythms	8
Hypertonia	7

After Startsev, V.G. (1976) *Primate Models of Human Neurogenic Disorders*, Hillsdale, N.J., Erlbaum.

'Voodoo deaths' in wild rats after restraint are described by Richter (1957). Upon release into a water jar they drown in a few minutes instead of surviving for the normal several hours.

Spontaneous behavioural abnormalities in animals are distortions from normal animal behaviours, not just analogues of psychopathologies in humans. Startsev (1976) observed convulsive attacks, paresis, paralysis, hyperkinesis and other motor disturbances in more than thirty baboons at the Soviet primate research laboratories at Sukhumi. The disturbances, further detailed in Table 1.5, were spontaneous inasmuch as they were not intended by the maintenance and experimental procedures used with the animals. Seventeen other baboons were thus employed in deliberate attempts to produce experimental models of these spontaneous baboon disorders, and the symptoms were found to result from irregular disturbances in the animals' living routines on the one hand, and forced physical restraint on the other. In all cases, convulsive attacks and other 'hysterical' motor disturbances were associated with stressful situations accompanied by intense motor activity, such as attempting to flee from a handler or struggling against physical restraint.

In pets

When experimental animal abnormalities model spontaneous

animal abnormalities it is safe to look for common originating situations, and it is likely that the spontaneous motor disturbances in Startsev's colony of baboons originated in the same way as those of his experimental animals. When originating situations that disable animals also disable humans, such as those depicted in Table 1.6, it is reasonable to hope that remedial techniques applicable to animals will also have some applicability to humans. In the following case, a therapy devised with humans (Wolpe, 1958) is applied to a phobic response in an animal. The case is described by Tuber, Hothersall and Voith (1974) who have proposed a clinical psychology of animals for dealing with spontaneous abnormal behaviours in animals, particularly pets. They illustrate their proposals with cases where psychological principles and procedures derived from normal animal research served to alleviate abnormal animal distress. Higgins is one such case.

Table 1.6 Comparative factors predisposing hysterical and motor disorders in humans and captive baboons

Humans	*Baboons*
Overprotection	Social deprivation
Psychological trauma	Conflicting social relations
Reaction to physical trauma	Reaction to aggression
Exhausting physical labour	Resistance to immobilization
Chilling	Chilling
Requirements of adolescence	Sexual and social life changes
Imitation	Imitation
Voluntary hysterical symptoms	Conditioned hysterical attacks

After Startsev, V.G. (1976) *Primate Models of Human Neurogenic Disorders*, Hillsdale, N.J., Erlbaum.

HIGGINS

Higgins is an affable four-year-old English Sheep Dog of Goliath proportions whose tranquil demeanor was breached only by an intense fear of thunderstorms. At the first indication of an impending storm, Higgins would begin an accelerating pattern of aimless pacing, profuse salivation and marked panting which was rapidly climaxed by the hurtling

> of his 110-pound body against any obstacle in a futile attempt to escape. (Tuber, Hothersall and Voith, 1974, p. 763)

Higgins was cured by the method of desensitization by counter-conditioning developed from Wolpe's (1958) studies of experimental neurosis in cats. Tuber, Hothersall and Voith prepared a stereophonic reproduction of a thunderstorm and played it to Higgins at increasingly louder intensities as he learned to tolerate each intensity level without any signs of fear. As they describe it:

> Training was initiated in the laboratory in daily sessions lasting one hour. The beginning intensity for each session was always slightly less than the terminal level achieved during the preceding session. . . . We had . . . progressed from a thunder intensity level of a meager 35 decibels to that of a resounding 75 decibels when a typical summer thunderstorm intervened to test our efforts. Happily, the owner reported that Higgins initially exhibited only a mild version of the original fear. (Tuber, Hothersall and Voith, 1974, pp. 763–4)

Additional cases of fears in companion dogs are described by Hothersall and Tuber (1979).

Animal contributions to medical science

Animal models in physical medicine

Several of the cases described above and in later chapters bear marked similarities to human responses to similar situations and incidents – paralysis under extreme fear, destructive behaviour at losing a sexual partner, perceptual disorientation during prolonged isolation, and thunder and lightning phobias. Their study may therefore be profitable not only for the alleviation of unintended animal suffering but also for the light it may shed on human problems, that is, as models of human mental illnesses.

In physical medicine, studies with animals for the benefit of humans are commonplace, and Gay (1967) has vigorously argued for further development of comparative medicine for the combined sakes of humans and animals. Leader (1967) illustrates how the study of anthrax and other natural animal diseases have contributed to human and animal welfare, sometimes in unexpected ways, and a document issued by the American Public Health Association (1967) lists benefits to animal and human health derived from animal studies in the areas of nutrition, surgical techniques, drug therapies and vaccine production. Beyond these, animal studies contribute to knowledge of the basic physiology of pathological states.

Among the nutritional benefits are the identification of the role of cholesterol in cardiovascular disorders, of the relevance of exercise in the prevention of myocardial infarction, and the value of fluoridation and vitamin therapy for deficiency diseases.

> In fact, except for ascorbic acid, thiamin, niacin, and Vitamin D . . . the search for the identification of other vitamins and minerals was motivated largely by work done in agricultural colleges on experimental animals. (American Public Health Association, 1967, p. 1598)

Animal experimentation also contributes to increasing the world's food supplies, both from the standpoint of raising food production by attention to farm animal welfare (Kilgour, 1978), and to the need for developing improved nutritional foodstuffs. In this respect:

> In addition to animal researches in nutrition providing better food for man, they have provided better food for animals. In fact, entire new industries of considerable economic importance have been developed in animal nutrition as, for example, the dog food industry and even a sizable industry providing food for cats (McCann and Stare, 1967, p. 1603)

With respect to surgical techniques, Winterscheid (1967) details contributions of animal preparations for the understanding of haemorrhagic shock ('a state of depressed consciousness; pale, cool, moist skin; increased respiratory rate, and a thin weak fast pulse') and for cardiac surgery and organ transplants: 'Thus . . . children and adults whose lives have been incapacitated by cardiovascular lesions may now . . . expect to have such defects corrected.' And concerning drugs, therapeutic actions of antibiotics and sulfonamides for infectious disease, antimalarials and others for parasitic diseases, diuretics for kidney disorder cases, as well as drugs to combat arthritic conditions, which occur in horses, dogs and cats, as well as humans, were all discovered or developed through investigations using laboratory animals (Robinson, 1967).

Animal models in psychological medicine

Animals as replacement

Perhaps it is in psychiatry that the most dramatic recent advances in drug therapies have occurred. For neuroses and mild emotional disturbances the use of minor tranquilizers has reached a point of abuse, and animal studies are required to discover the modes of action and addicting potential of such drugs. As for more serious, hospitalized, patients, the mental hospital patient population in the United States has declined by about two-thirds since antipsychotic drugs were introduced. The principal drugs employed in psychiatric institutions are phenothiazines (e.g. chlorpromazine) with schizophrenics, monoamine oxidase inhibitors (e.g. iproniazid) and tricyclic compounds (e.g imipramine) with depression, and lithium salts with mania. In many cases the psychiatric value of such drugs was discovered accidentally, and animal studies are now conducted for the discovery and classification of superior compounds, and also for the determination of the pharmacological basis of psychosis (see Chapter 7).

In addition to the above, the establishment of animal models is necessary for the study of side-effects attributed to psychiatric drugs. One such side-effect is tardive dyskinesia, which is a condition of 'protruding, twisting and curling movements of the tongue; pouting, sucking, or twisting lip movements; bulging of the cheeks and various forms of chewing movements' (Tarsy and Baldessarini, 1976, p. 30) that appears after prolonged phenothiazine treatment. To combat the problem, numerous studies of tardive dyskinesia with mental hospital patients have been reported, often with no mention of consent by the patients or by relatives. Such studies violate ethical principles of *valid consent* (the subject properly understands the experiment) and of *prohibited subjects* (experiments should not be performed on the mentally sick, the aged, or the dying) proposed by Pappworth (1967) for medical experiments with humans, so the problem must be solved through animal experimentation of the kind reported by Weiss and Santelli (1978). These workers employed two monkeys as models for the investigation of tardive dyskinesia and obtained similar results to those obtained clinically with humans. Figure 1.1 shows one of the monkeys after three months' administration of weekly doses (0.25 mg/kg at first, then 0.50 mg/kg) of haloperidol in fruit juice. With the lower dose, dyskinesias in the forms of tongue and perioral movements and twitching, crouching and writhing appeared after 9 weeks on the drug regime, only to disappear with two further weeks of treatment. When the drug dose was raised, however, as is sometimes the case with humans, the original dyskinesias returned in exaggerated forms. Such effects might be acceptable in severe human clinical conditions that respond only to pharmacological treatment but they are ethically unacceptable for human medical research. With drugs, especially, Pappworth proposes a third ethical principle, the 'principle of previous animal experimentation'. These ethical principles of Pappworth underpin much of the

Figure 1.1 Monkey 42 after the 90th weekly dose of haloperidol (0.5 mg/kg). During this episode the animal grimaced, yawned, twisted into unusual postures, and, as shown, protruded its tongue as it made incipient chewing movements. Note the slight ptosis of the left eyelid and the peculiar clasped position of the front paws. (From Weiss, B., and Santelli, S. (1979) 'Dyskinesias evoked in monkeys by weekly administrations of haloperidol', *Science*, 200, 799–801, Copyright 1979 by the American Association for the Advancement of Science. Reprinted by permission of the author.)

experimental work performed on animal replacement for humans. (Ethical principles for animal experimentation are discussed in Chapter 2.)

There are many contributions of experiments with animal replacements for humans in the study of psychopathology, including those on symptomatic equivalences, fundamental processes, reactions to environmental stresses, and attitudes to treatment. Concerning primates in particular, Mason (1968) provides three distinct perspectives for experiments: from an evolutionary comparative standpoint to enlighten ourselves about human disabilities through comparison with animal similarities and differences; for pure research out of direct concern about the animals; and as permissible substitutes when humans cannot ethically be employed for examining simple, specific behavioural functions. From the third perspective, which generalizes Pappworth's principle of prior experimentation, there are certain varieties of experimental procedures possible with lower animals that would not be permitted with humans; screening for dangerous drugs and the creation of 'experimental neurosis' are obvious examples. Others are experiments that have focused on selective breeding for obesity in mice, on susceptibility to ulceration in rats, on nervous instability in dogs, and on addiction to opiates in monkeys. Experiments on decorticate and decerebrate animals for the study of brain functions also cannot be done on humans, nor can many of the experiments in medical laboratories, including some of those described above. In these cases, ethical and rational considerations must be balanced against whatever benefits the experiments may have. These matters are addressed below in Chapter 2.

Animals and mental illness

As Henri Ey (1964) has said:

> Le concept de 'Zoopsychiatrie', representerait bien ce scandale logique et moral que certain 'cartesians' seraient enclins a denoncer, si n'etait acceptée depuis longtemps l'idée qu'il y a une 'Psychologie animale' dont l'objet et le psychisme ou, si l'on vent, la *psychoide* animale.

Even if normal animal psychology is accepted, not in the literal sense of psyche-ology but in the sense of the scientific study of behaviour, its possible contribution to human psychopathology is not obvious. Nevertheless animals as well as humans display a variety of abnormal behaviours, some of them stressful and disturbing. When these abnormalities are created experimentally, discovery of the relation of cause to effect is possible in principle, but when behavioural abnormalities occur spontaneously in nature their origins and sustaining sources can only be surmised. One such supposition was the medieval belief in demoniacal possession, and another is the modern belief in mental disease. The question of possession of animals by devils has not been debated for centuries (Evans, 1906), but there is a question of mental illness in animals, and the question is said to be unresolved (Zubin and Hunt, 1967). Resolution of the question depends on what mental illness is intended to mean.

ILLNESS AS CAUSE The question of the existence of mental illness in animals implies its existence in humans, and the answer depends on the status of the assertion with humans and the characteristics that differentiate humans from lower animals. Frequently, mental illness is taken to mean the cause of abnormal behaviour. This is the sense in which behavioural abnormalities are diagnosed as symptoms of psychiatric diseases. However, as a cause of abnormal behaviour mental illness, like demoniacal possession, is not an empirical discovery but an assumption embedded in a set of fashionable beliefs. Its existence in either animals or humans cannot be discovered by observation or experiment because it is not a causal agent like a toxin or a germ. It is mistaken to assert that mental illness is the cause of abnormal behaviour, animal or human, because the only evidence of mental illness is the behaviour it is supposed to cause. In this sense mental illness does not exist in animals because it does not exist in humans.

ILLNESS AS DESCRIPTION However, in the sense of describing or categorizing the disturbing and distressing behaviour frequently seen by clinical psychologists and psychiatrists, mental illness may be said to be demonstrable in humans. In this case it is an effect (clinical picture), not a cause, and psychopathology is a possible synonym for extraordinary distress so long as its status as synonym, not determinant, is not overlooked. In this descriptive usage, mental illness is a private experience of the organism in distress, and can only be inferred by an observer from public verbal or non-verbal behaviour of the disturbed individual. Animals cannot say if they are mentally ill or well but this is not a disadvantage unique to animals, for neither can many humans. Claims by psychiatrically diagnosed psychotics that they are actually sane are often disregarded as mistaken, and pleas of insanity by legally convicted criminals are frequently contested by prosecuting lawyers and expert psychiatric witnesses. Thus mental illness may be attributed with doubt or confidence to animals and humans alike according to how public signs of private distress are interpreted.

Qualification of illness in this case by ‘mental’ serves the useful purpose of separating statistically normal from unusual responses to private events, but the practice is dangerous because the qualification is easily mistaken for a separation of psychical from physical sources of distress. Distress responses with organic origins must be differentiated from those without them, but the difference is not physical-organic versus psychic. Physical and mental illness (in the descriptive sense) both have physical origins; organic damage in the one case and environmental contact with behaviour in the other. All the examples of animal behavioural abnormalities described earlier are physical behaviours established and maintained by physical environments. They are not instances of illnesses *of* the mind or pathologies *of* the psyche; they are instances of disordered behaviours that could have distressing accompaniments. Mental illnesses apply to the accompaniments not to the causes of disordered public behaviours.

ANIMALS AS MODELS Finally, the question of mental illness in animals pertains to the status of abnormal animal behaviours as models of mental diseases in humans. In 1960, P.L. Broadhurst thoroughly reviewed early Russian and American descriptions of experimental neurosis and concluded that 'no animal analogue of neurosis as a disease entity had been demonstrated', a conclusion he repeated in 1973. Broadhurst's negative answer in 1960 applied to the existence of neurosis *as a disease entity*, not the existence of behavioural disorders in animals. That answer is not now surprising because 'the terms in which the question was put have altered'.

The old terms took a disease model of human neurosis for granted and looked for structural equivalence in animal and human disorders. It now seems that aspirations for structural equivalence were never realistic, and even though Chamove, Eysenck and Harlow (1972) report comparable personality factors in macaque monkeys and normal humans it does not follow that abnormal personality structures of monkeys and humans will be identical. Even normal apes, which in some ways mentally represent the world like normal humans (Mason, 1976), obviously do not think in human terms, so how could an abnormal thinker like an obsessional neurotic or a paranoid schizophrenic be modelled by an abnormal ape? The answer may be found by way of a functional analysis of the public elements of obsessional and paranoid behaviours.

The functional approach differentiates between behavioural disorders and unlawful behaviour, and dictates functional analysis of behavioural disorders on the basis of normal behavioural laws. Etiology, prognosis and therapy are equivalent to the development, maintenance and modification of behaviour, and Ullman and Krasner (1969) claim that:

> Abnormal behavior is no different from normal behavior in its development, its maintenance, or the manner in which it may eventually be changed. The difference between normal and abnormal behaviour is not intrinsic: it lies in societal reactions to them. It is more fruitful to ask how a person develops *any* belief than how he develops a false belief. The principles of the development of 'proper' beliefs are the same

> as the principles of the development of 'false' beliefs. (Ullman and Krasner, 1969, p. 92)

Arguably there are intrinsic differences between some normal and abnormal behaviours, but the argument need not revive the dogma of mental illness in its causal form. The attribution of mental disease to humans has led to a futile search for its nature in humans and also, as Broadhurst (1960) discovered, in animals. The question of causal mental illnesses in animals as models of causal mental illnesses in humans is not a question that is open; it is a question that is mistaken. The question is whether laws of human behaviour are visible in animals, and the answer of modern experimental psychology is 'Yes.'

Claims for animal–human similarities in pathological processes or outcomes encompass psychoanalysis, psychiatry and ethology as well as experimental psychology. From ethology, Hinde (1962) analyses the relevance of animal studies to human neuroses in three stages:

1 Immediate responses to stress;
2 How these responses become habitual and distorted;
3 How they come to occur in new contexts;

and lists as discoveries from ethological and laboratory investigations: approach–avoidance reactions, displacement, sexual inversions, regression and tonic immobility as stress reactions; sensitive periods and early experience as fixation mechanisms; and superstitious behaviour and experimental neurosis as examples of transference to novel contexts. Similarly, from psychiatry, Jones (1971) cites analogies between animal and human behaviours in the areas of stereotypies, aggression, conversion hysteria, attention-seeking hysteria, sexual behaviour, and separation syndromes. Concerning this latter, Jones and Barraclough (1978) relate self-aggression (auto-mutilation) observed in mammals raised in partial isolation to the self-mutilation – 'scratching, biting, hair pulling and head banging' – that human children sometimes show. With infants, Bowlby's (1976) list of animal contributions to the hospitalism syndrome is given in Chapter 8.

Similarities between animal and human psychologies may be examined with respect to origination and symptomatology, and from functional and structural perspectives. Similarities in functional processes between animal species, at least at the level of the law of effect, are generally acknowledged by psychologists, and B.F. Skinner, in particular, has constructed a psychology of humanity on the basis of experiments with animals. Desmond Morris and Konrad Lorenz have done much the same from an evolutionary point of view, especially stressing animal–human parallels in the areas of aggression, courtship and sex. It is work of this nature that led to the Nobel awards to von Frisch, Lorenz and Tinbergen mentioned at the beginning of this chapter.

Animals and diagnosis

Ellenberger (1960) illustrates a value of animal studies for psychiatric diagnosis. He distinguishes between the disease processes for which a human patient may be hospitalized in the first place, and the reactions of the patient to the hospital after admission. These he likens to behaviour syndromes exhibited by animals confined in zoos. Unlike Seligman (1975), who claims a parallel between animal and human depression as a disease process, or Ferster (1966), who proposes similarities in environmental causes of retarded behaviours in animals and humans, Ellenberger draws a parallel between animal and human reactions to a common environment – confinement, or loss of individual freedom.

Four animal–human parallels in confinement syndromes are suggested. For zoo animals, Ellenberger lists:

1 *Captivity trauma*, which appears in the form of acute depression, prolonged stupor and refusal to eat.
2 *Nestling*, or the establishment of the cage as a home, and the demarcation of a personal territory within it.
3 *Social competition and frustration*, expressed as dominance competitions and age and sex conflict interactions.
4 *Emotional deterioration*, in the form of repetitive stereotypies like pacing, rocking and swaying, coprophagia and faeces smearing, and signs of apathy and depression.

The comparable syndromes in humans are:

1 Anger, withdrawal and negativism as the human captivity trauma reaction.
2 Nestling, with seating and eating locations as human territories and the hospital ward as the home.
3 Frustration manifested as pettiness and interpersonal jealousies.
4 'Alienization' in the form of infantile regression, bizarre catatonic gestures, aggression, agitation and delusions, and infantile behaviour characteristics as equivalent to emotional deterioration in animals.

This parallelism is of value because it differentiates reactions to hospitalization (which would affect a psychiatric diagnosis made in an institution) from behavioural manifestations responsible for hospitalization, upon which an original diagnosis would be made. To the extent that diagnosis is a desirable psychiatric enterprise then this is an essential differentiation for the improvement of diagnostic reliability, but more importantly still it reinforces the pleas of many recovered patients for reform in asylum management. From the time of John Perceval, son of a British Prime Minister (Bateson, 1961), to that of the New Zealand novelist, Janet Frame (1961), at least, indignities suffered by mental patients have been vividly but ineffectively described. It would make a fitting irony if the parallel drawn by Ellenberger between animal and human responses to confinement were to succeed in effecting reforms in mental hospital management where autobiographical and fictionalized accounts by articulate writers of patients' humiliations have mostly failed.

Animals and treatment: therapeutics, prosthesis and prevention

In physical medicine, treatment by therapy aims to cure an unhealthy organ, whereas prosthetic treatment replaces or bolsters an imperfect organ with an artificial aid. Prosthesis remedies a disability without affecting the disabled organ; therapeutics eliminates disease. Thus, bone-setting and surgery are therapeutics while wooden legs and wheelchairs are prosthetic devices. Supportive drugs like phenothiazines and benzodiazapines used in psychiatry are more prosthetic than they are therapeutic.

The medical model of mental illness is almost entirely oriented to therapeutics, so that even critics of the disease model of psychopathology propose only alternative methods for cure. Behaviour therapy is substituted for psychotherapy, but treatment still is aimed at cure. Yet if behaviour pathologies are maintained by abnormal interpersonal transactions it is not the cure of an intrapsychic malfunction that would be the objective of treatment but the prosthetic rearrangement of the disabling transactions.

Among experimental psychologists, Lindsley (1964a; 1964b) has emphasized the importance of prosthetic environments in education and geriatrics, and Ayllon (1974) has defended the requirement of total control of a hospital environment for the maintenance of non-psychotic behaviour. The defense was in response to a challenge by Sherwood and Gray (1974) that some classic cases of behaviour modification treated by Ayllon and his colleagues had relapsed. While such a challenge is meaningful in the context of a medical cure model for mental illness, it is out of place with reference to a prosthesis approach to treatment, because only in a proper supporting environment can normal behaviour be maintained, and normal mental hospitals are arranged to support abnormal, not normal, behaviour, according to Rosenhan (1973) and others.

In psychological medicine the dividing line between treatment and prosthesis is a fine one. Consider the hypothetical illustration of meanings of abnormality by Ferguson (1968).

> The term 'abnormal' in relation to the behavior of domestic birds is susceptible to different usages by husbandrymen, veterinary clinicians, and behaviorists. Thus the commercial egg producer who has learned to expect consistency of performance from commercial laying strains, producing in excess of 230 eggs per annum, may insist on classifying as alarmingly abnormal any primitive behavior, such as marked broodiness, which seriously compromises egg production targets. The behaviorist accords greater respect to the biological significance of this normal maternal trait and may be inclined to regard the degree of suppression of broodiness, sought for in commercial layers, as prejudicial to preservation of the species and, therefore, classifiable as abnormal. Similarly, increased exploratory pecking activity . . . shown to occur in calcium-deprived birds, is normal by a behaviorist's definition, since it has a sound biological purpose aimed at the eventual recognition of the corrective nutrient and thus the preservation of life. The same activity, since it represents a deviation from that of birds receiving adequate calcium, would appear to the clinician as an abnormal behavior resulting from the abnormal nutritional status. (Ferguson, 1968, p. 190)

With an abundant supply of calcium the bird is essentially normal; without it, the bird is a compulsive pecker. From one view the bird is cured by a ration of calcium and relapses when the supply is exhausted; from another, calcium is a piece of prosthetic equipment, like spectacles, that remedies a behavioural disability without repairing a defective organ.

At a more complicated behavioural level are the different effects of maternal separation on bonnet and pigtail macaque monkeys. In the bonnets, maternal loss produces minimal behavioural changes, whereas pigtail infants react to separation from their mothers with agitation, stereotyped postures, withdrawal and depression. The differences could be that relative to bonnets, pigtails are more prone to anaclitic depression, but it is more likely that the difference is the result of different mother–infant interactions in the two species. Bonnet mothers are permissive with their infants, who interact freely with other adults; pigtail mothers guard and restrain their infants from social intercourse. In the absence of their mothers, pigtail infants are social isolates, but bonnet infants soon find guardian 'aunts'. Such aunts provide bonnet infant monkeys with prosthetic environments that pigtail infants are denied.

A simple prosthetic apparatus for a baby rhesus monkey is a piece of soft material. Infant monkeys are difficult to rear in barren metal cages, but when an upright cone covered with terry-towelling is placed in the cage the animals easily survive and spend much of their time clinging to the cone. When given the choice of clinging to a cloth-covered cone ('cloth mother') or a bare-wire cone fitted with a feeding bottle ('wire mother'), infant monkeys cling mostly to the cloth mother, visiting the wire mother only briefly to feed. The terry cloth is an important prosthesis for the isolated developing infant monkey, but even more important is its function for preventive psychiatry. Monkeys raised in such a manner do not show the psychiatric disorders of adulthood found in monkeys reared in isolation, a finding that has had ample application to the paediatric care of orphaned human children.

2 Animal experiments and animal welfare

Animal suffering

The purpose of medical science is to alleviate suffering, human in the case of general medicine and animal in the case of veterinary medicine. Psychiatry is the specialized branch of general medicine that attends to suffering attributable to mental illnesses in humans, but there is no comparable specialization in veterinary medicine. Nevertheless, as we saw in Chapter 1, there are numerous animal analogues of human functional disorders, and there are also anomalous animal behaviours that will be listed later in this book. It is necessary to make some assessment of animal suffering in all of these cases, not only as a basis of therapeutics but also for minimizing suffering in experimental psychiatry.

Criteria for judging animal suffering

Some animal afflictions are laboratory creations (which raises the ethical questions discussed below), but many occur under accidental circumstances. In farm animals, for instance, tail biting in pigs (Colyer, 1970) and hysteria in hens (Hansen, 1976) are counterproductive results of complex interactions between space, temperature, humidity, nutrition and population density conditions. These and other disturbances in farm animals' behaviour have fostered the scientific study of animal welfare on farms, from which Dawkins (1980) has derived six criteria for judging suffering in animals.

1 What are the conditions under which the animals are kept?
2 Are the animals physically healthy?
3 Does the behaviour, physiology and general appearance of the animals differ from that of genetically similar animals in less restricted conditions?

4 Is there evidence of severe physiological disturbances?
5 What is the cause of the behavioural differences established under 3?
6 What conditions do the animals prefer?

These criteria might be applied to zoos and laboratories and other circumstances wherein animals are subject to human control, although the pica and stereotypies frequently displayed by zoo animals would not constitute suffering by most of Dawkins' criteria, nor would the bizarre behaviour of some chicks that were used to examine the effects of learning and maturation on development. The chicks were taken directly from the incubator to a warm well-lit 60 × 50 × 30 cm isolation box where they were watered and fed without ever seeing the keeper or an older chicken.

> After an isolation of no more than a week [they] . . . started passionately chasing flies. . . . Sometimes they succeeded in catching one, but equally often all their efforts were in vain. *As if possessed*, they ran behind the fly aimed at and went on chasing as if after a phantom when it escaped them. (Katz, 1937, pp. 216–17, italics added)

As with zoo-confined animals, some chicks also emitted stereotyped ritualistic movements uncharacteristic of normally reared chicks. Such side-effects of an experimental procedure, which also occur with isolated humans, serve to alert investigators to contaminating effects in laboratory studies. Although it is impossible to judge if there was suffering by the chicks, there cannot be much doubt in the following cases, which are examples of abnormalities in laboratory animals accidentally caused by routine caretaking procedures unconnected with formal experimentation. As a result of these cases, caretaking routines have been modified to reduce the risk of similar incidents in the future.

Animal distress in unexpected situations

Cupid and Dennis were primates used respectively in the laboratories of Tinklepaugh (1928) and Ferster (1966) for psychological investigations with nothing to do with distress. Both animals, however, exhibited highly distressed behaviour when their normal mates were replaced by substitutes in the course of investigations.

Cupid was a young male rhesus macaque monkey originally caged with an older female named Psyche. Some time later Psyche was removed and Topsy, another female, took her place. Cupid's response was not anticipated by Tinklepaugh, who wrote:

> Cupid eyed her from a far corner of his cage. Topsy strolled nervously about the cage . . . with her tail up so that her genitals were exposed to him. After about a minute of this behavior Cupid leaped upon her, seized her by the small of the back and hurled her across the cage.

Some months after this, with Psyche, Topsy and another female intermittently in and out of Cupid's cage, on an occasion with only Topsy present, Cupid began 'biting his hind feet much as he had formerly done in play'. On account of this he was taken for medical attention, on the way passing Psyche's cage.

> He looked back at the cage where Topsy still remained and then toward the cage containing the other two females. Psyche, who had seemed to be much upset by Topsy's presence in Cupid's cage, was now on the side of her cage, shrieking threateningly across at the other female. Suddenly, and with no previous sign of anger or particular emotion, Cupid lurched to the end of his chain and began to bite himself. In a few seconds, he tore huge jagged rents in his . . . legs . . . a three inch gash in his hip, ripped his scrotum open . . . and mutilated the end of his tail. (p. 230)

A comparable, though less dramatic, event occurred with Ferster's chimpanzee, Dennis. It happened during a long-term experiment involving Dennis and a female companion in a large complex multi-unit experimental space. At one point the female was replaced by another, whereupon Dennis pummelled, kicked and pushed her, and for several weeks went partially off his food. 'We could,' Ferster observed, 'have described Dennis as angry and depressed, and we would not have been too far off the mark' (Ferster, 1966, p. 55).

It is not difficult to find reasons for agreeing with Ferster's belief in Dennis's suffering, and for thinking that Cupid suffered also. These reasons are by analogy with ourselves: *analogy by response*, as when an animal squeals or limps; and *analogy by stimulus*, when we imagine ourselves in the same situation. We might confidently employ these analogies in assessing suffering in Dennis and Cupid, but there are many opportunities for error when the animals, like Katz's 'possessed' chicks, resemble humans less closely than do chimpanzees or monkeys (Dawkins, 1980).

Other examples of spontaneous behavioural signs of distress in primates kept for experimental purposes are reported by Startsev (1976). The incidents involved the removal of some juvenile hamadryas baboons from an indoor to an outdoor larger cage, where the juveniles were exposed to threats from adult males in an adjoining enclosure. Two animals 'showed a gait disturbance characterized by incoordination and a posture with the knees half flexed' which in one, the case of Zagreb described in Chapter 1, resulted in death with no evidence of organic impairment. Startsev (1976) describes another case.

> A hysterical motor disorder was produced in the baboon Azov under similar circumstances. He was transferred from a small home cage to a larger cage, which contained a female and was located within view of several other mature males. He had shown brief spasmodic attacks previously under the stress of being driven from one side of the home cage to the other; but now he developed frequent and prolonged convulsive attacks with paresis and paralysis of quite a different character. (Startsev, 1976, p. 135)

The disorders of Cupid, Dennis and Azov were not results of experimental procedures as such, but of incidental care and maintenance procedures. Similar problems can arise when cages and cagemates are routinely rearranged in zoos (Stout and Snyder, 1969). The case of the isolation-reared chicks is different, however, because it is an example of an experimentally obtained effect, even though the effect was not a pre-planned aspect of the experiment. The classic prototypical accounts of experimental neurosis discussed in Chapter 5 are similar, for the experiments of Yerofeeva and Shenger-Krestovnikova were about stimulus generalization and discrimination, not attempts to create animal models of human neuroses.

Laboratory accidents, household accidents, and traumatic events in the wild can all produce similar reactions in animals. The case of the wild chimpanzee, Merlin, was described in Chapter 1, and that of the home-reared chimpanzee, Lucy, appears in Chapter 8. Both cases show disturbances of posture that Mason (1968) includes in the primate deprivation syndrome and that are identifiable in human autistic children. A major characteristic of such children is head-banging, a behaviour that occurred in a male rhesus monkey following a laboratory misadventure (Levison, 1970).

> When the subject was 1 year old the experimenter began training him to enter a transfer cage. . . . However . . . the door of the transfer cage accidentally dropped on the subject, glancing off his head and shoulders . . . his response was to race away from the transfer cage and crouch in the left rear corner of the cage, where he sat huddled in the corner. After this the introduction of the transfer cage into his home cage

> was correlated with . . . refusal to enter . . . crouching, rocking and headbanging. He would sit in the left rear corner of the cage and rhythmically bang his head against the plastic wall of the cage.

Chapter 8 contains a further account of the circumstances of this animal which, were it human, would be a clear candidate for psychiatric or psychological treatment.

Apart from farm, zoo and experimental animals, ordinary household pets also show spontaneous behavioural disorders. The case of Higgins was described in Chapter 1. Hothersall and Tuber (1979) give other examples, including fear of thunderstorms by a German shepherd called Cindy and a Labrador retriever named Major. In Major's case there was a possible origin for the fear because an arc welder once exploded on a bench where he was chained, but in the other case, no ready explanation of the fear was available.

> Cindy was a gentle 4-year-old German Shepherd, who was acutely afraid of sudden loud noises and of thunderstorms. Long before the storm became apparent Cindy would begin to pant, whine and pace. As the storm became imminent, she became increasingly disturbed and would begin to discharge from the nose and mouth; each thunderclap would elicit strong and uncontrollable trembling [ending in] a final collapse into a flaccid trancelike state after which she would remain completely unresponsive for periods up to 24 hours. (Hothersall and Tuber, 1979, p. 247)

These illustrations are sufficient to suggest, at least, that lower animals can experience suffering parallel to that of humans, and that the situations that occasion it are similar in both cases. They document the need for psychiatric investigations of animals not only for their possible relevance to the understanding of human aberrant behaviour but also for their practical possibilities for alleviating animal suffering that stems from intentional or unintentional sources. Naturally, such investigations cannot proceed on the basis of nineteenth century anthropocentric psychology, in which 'it is no worse from an ethical point of view to flay the forearm of an ape or lacerate the leg of a dog than to rip open the sleeve of a coat or mend a pair of pantaloons' (Evans, 1898, p. 99). Nowadays this is an unthinkable opinion and considerable attention is paid to the questions of ethical (Marcuse and Pear, 1979; Sechzer, 1981) and humane (Russell and Burch, 1959) procedures in experiments with animal subjects. At the heart of these questions is the one of correspondence in animal and human natures.

Ethics and animal experimentation

Animal nature and human nature

If human psychopathology is to be studied by way of animal psychopathology, then some correspondence between animal and human natures must be evident. A few animal–human similarities have been illustrated above, and elaborate generalizations from animal to human nature in the areas of territoriality, aggression and sex are claimed in popular works by Ardrey (1978), Lorenz (1966) and Morris (1967). Walker (1982) has convincingly documented by way of comparative brain anatomy the steps from animal to human thought, and Thorpe (1974) shows several ways in which animal natures are similar to those of humans – animals too can learn, plan ahead, form concepts, use tools, communicate with each other, count, and show visual and musical appreciation. Griffin (1976) uses several of these as criteria of consciousness in animals, and Harlow, Gluck and Suomi (1972) give examples of intellectual, motivational and pathological generalizations between human and non-human animals. 'There is', they claim, 'only one way to test the limits of interspecies generalizations and that is by experiment.' At the same time, however, the authors recognize that evaluation of such generalizations 'may be more an art than a science.' The essence of the art is identified by Frey (1976).

> Generally speaking, a statement about similarity reflects an experimenter's claim that investigated phenomena have common characteristics. The term similarity also indicates that the experimenter has observed characteristics in which these phenomena differed. In classifying phenomena as being similar, however, the experimenter implies that he considers the differences he observed as negligible. (Frey, 1976, p. 7)

For some, the difference between humans and animals is not negligible. Bannister (1981), for instance, arguing that animal experimentation in psychology is a fallacy, acknowledges that animal–human similarities exist but cites economic and ethical reasons as the principal justifications for animal studies. He attacks reductionism and naive realism as the bases of scientific psychology, and rejects the image of psychology as a science under the influence of Darwinian biology. To Bannister, psychology as a study of behaviour is the study of man as an animal. Against this he stresses the uniqueness of humankind and argues for psychology to be the study of experience – the study of human experience alone. This itself is an economic argument if resources for the study of both are unavailable, and is anthropocentric in ignoring the possibilities, first, that a psychology of animal experience could exist and, second, that experimental psychologists could be interested in animal suffering.

Beyond that, as the study of human experience, psychology is improperly and unnecessarily restricted. Experience can only be communicated by means of public behaviour, so at least psychology must concern itself with human behaviour as well as with human experience, and human experience is normally expressed through verbal behaviour acquired via the medium of inherited gestures, some of which humans and animals share (Darwin, 1872).

Human nature is anything but clearly understood on the basis of experiment or of self-examination by humans. In a recent symposium organized by the British Psychological Society on Models of Man (Chapman and Jones, 1980), Jahoda asks the question, 'One model of man or many?' and answers, 'I have come to only one firm conclusion: a rejection of the idea of one unitary model of man for psychology. I do not believe that one should or even could strive for it.' And by the same token, there is no reason to demand a unitary animal model to cover all varieties of human psychopathology, or to assume that animal psychology cannot be generalized to humans. The examples I have given already, strongly suggest that it can.

Scientific and moral judgment

Apart from arguments against animal experiments in psychology on the grounds of dogma there are also ethical reasons given against experimentation with animals. These usually consist of reasons for not conducting experiments, but they may take the more positive form of guidelines or directives for the employment of certain experimental precautions. Precautions and guidelines are necessary because non-human animals cannot control their degree of participation in experiments and so must be protected by ethical principles and procedures adopted by experimenters. Thus there are two kinds of questions involved in the use of animal subjects in psychiatric or psychological research for the benefit of humans. One pertains to the validity of the results for humans – the rational, or result generalization, reason; the other pertains to the ethics of data collection – what can be done experimentally with animals that cannot be done with humans. The ethical question arises only if the rational question is answered affirmatively first, for if the results do not generalize then there is no reason to conduct the experiment. If the rational question is answered affirmatively then the ethical question only arises when there is opposition between equally desirable values; the need not to cause suffering in animals and the need not to allow suffering in humans.

The essence of the ethical problem can be expressed by what I call Singer's contradiction and its Wisconsin converse.

SINGER'S CONTRADICTION

> The researcher's central contradiction exists in an especially acute form in psychology: either the animal is not like us, in which case there is no reason for performing the experiment; or else the animal is like us, in which case we ought not to perform an experiment on the animal which would be considered outrageous if performed on one of us. (Singer, 1975, p. 49)

THE WISCONSIN CONVERSE

> There is only one way to test the limits of interspecies generalizations and that is by experimentation . . . . If nonhuman data do not generalize to data derived from human beings, can human data be used to predict within reason the supposedly homologous behaviour of nonhuman animals? (Harlow, Gluck and Suomi, 1972, p. 709)

How, that is, can we know how much an animal is like us without experimentation, and how, without knowing how much an animal is like us, can we judge when the animal is suffering? By analogy with ourselves, argues Dawkins (1980), and the analogies are discovered by experiment.

Singer's contradiction and the Wisconsin converse argue logical dilemmas. They differ from the personal dilemma of Davis (1981) who laments that he must sometimes conduct experiments with aversive stimulation of animals that he likes to keep as pets. He writes:

> I have run experiments using aversive stimuli since 1966. In both my training and my own research the use of animals has been axiomatic. On the other hand, I have an abiding fondness for animals: not simply for conventional pets, but for the kind of animals that one is likely to use as an experimental psychologist. Over the past five years I've trapped, fed and released as many as 200 mice who managed to find their way into my house. I've had the rare opportunity to rear a red squirrel from weanling to mature adult, and currently have a domesticated pet rat who, on most days, has the run of the house. (Davis, 1981, p. 63)

At the risk of exaggeration, the Davis dilemma might be compared to that of a hangman. Bassford (1981), explaining role differentiation in the ethics of psychological research, begins:

> If you or I were to put a noose around someone's neck and then hang them, we would rightly be considered moral monsters. But if the royal hangman were to do this while performing his official duties, his action would *not* be morally culpable. Indeed, until recent times, it would be morally laudable. (Bassford, 1981, p.27)

Thus, as a scientist in an officially approved educational institution, Davis could appeal to a duty assigned to him by society as a reason for conducting an experiment that might be questionable to him morally as a private citizen. (In this respect, Reed (1981) makes a useful distinction between personal moral standards and codified ethical principles: 'Moral behavior is the *practice* of virtue .... Ethics is concerned with abstract moral principles and their codification.')

So Davis could resolve his personal dilemma by changing his profession from psychology to law or moral philosophy, just as the hangman is free to change his. However, this still leaves society with the same dilemma as it faces with hangmen: does it require practitioners of the profession, or does it not? And if so, to what extent? The answer cannot come from illustrations of experimental successes or failures, for these are post- not pre-experimental events, but from the relative orders of priority that society gives to competing values.

Diamond (1981) expresses the Davis dilemma in the context of two extreme views on animal experimentation. The context is broader than Davis's lament, which concerns itself only with the infliction of pain, for Diamond considers the justification for using animals in experiments at all. According to the *first view*:

> Within certain limits, experimental animals may be regarded as delicate instruments, or as analogous to them, and are to be used efficiently and cared for properly, but no more than that is demanded. (Diamond, 1981 p. 341)

By this view, the standards of welfare for animals would be based on *scientific judgment* and scientific judgment alone. Possible cruelty to animals in experiments would be controlled by the disapproval of scientific colleagues on the one hand and the likelihood of producing unreliable experimental results on the other. Decisions would be based on the acquisition of knowledge as the ultimately desirable value for cultural survival. But, says Diamond, there is a *second view*, that:

> Within certain limits, animals may be regarded as sources of

> moral claims. These claims arise from their capacity for an independent life, or perhaps from their sentience, but in either case the moral position of animals is seen as having analogies with that of human beings. (Diamond, 1981, p. 341)

The second view is based not on scientific judgment but on *moral judgment*, for which the scientist *qua* scientist has no more expertise than anyone else. Criteria for moral judgments rest on contributions to the welfare of the powerless, concern for others, about which scientific judgments are in principle neutral. These first and second views, scientific and moral judgments, present the dilemma facing experimenters using animals, for between them the cultural values of desire for knowledge and concern about others are likely to come into conflict (Marcuse and Pear, 1979).

The category of judgment, scientific *versus* moral, establishes how the views differ, but, Diamond argues, there are matters common to both views hidden in the phrase 'within certain limits.' These are as follows.

AN ANIMAL'S LIFE IS LESS IMPORTANT THAN A HUMAN LIFE
Neither view is opposed to this and neither view would take the extreme position of nineteenth-century anthropocentric psychology. Less dramatically, however, a group of psychology students surveyed by Keehn (1982) demanded less strict requirements for experiments with animals than for experiments in which human subjects were involved, but the differences were not large. On a 5-point scale the statement 'The subject must be protected against any foreseeable injury' was rated 4.9 on average for humans and 4.0 for animals, and the largest human–animal difference was 4.4 for humans and 2.9 for animals on the statement, 'Drugs with unproven effects must not be employed in the experiment.'

The sample that produced these values may not be representative of society at large, but it is unlikely that any group of individuals would reverse the order of stringency between animal and human experimental subjects' rights. Perhaps the fairest modern expression of what the 'certain limits' are in the general population is Dawkins' remarks that:

> Governments are under pressure to change the laws on the treatment of animals. Scientists are on the defensive over their experiments on animals. Farmers are criticized. But most people go on eating animals, demanding that the products that they eat or wear are tested, wanting better drugs or transplants or vaccines to save their lives. (Dawkins, 1980, p. 11)

ANIMALS HAVE MORAL CLAIMS IN SOME CIRCUMSTANCES
Whereas the first area of agreement between *view one* and *view two* may represent a softening of the moralist's extremism of banning all use of animals for human ends, the second region of agreement represents a retreat from the extremism of nineteenth century anthropocentric psychology. The retreat takes two independent lines, one differentiating moral obligations to animals in experiments from moral obligations in normal life, the other distinguishing between types of experiments and the authorizations required for their performance. In the first of these cases, Table 2.1 shows four hypothetical individuals, A who concedes no moral obligations to animals at all, B who accepts moral obligations in normal life but not in experiments, C who insists that moral obligations are indispensable always, and D who would deny moral obligations to animals in normal life but treat them with respect in the course of experiments. Of these imaginary individuals, A would be the absolute representative of anthropocentric psychology, B could be the animal-lover but uncompromising scientist, C is the champion of animal rights in life and in experiments, and D is the scientist who only respects animals as instruments for research. In absolute categories, A, B, C, and D are ghosts, but they represent compass points from which the 'some circumstances' under which animal experiments can be performed can be approached.

Table 2.1 Categories of moral obligations

		In normal life	
		NO	*YES*
In Experiments	*NO*	A	B
	YES	D	C

Concerning the case where different experiments require different kinds of authorization, Ross (1978) describes a procedure adopted by Sweden. By this procedure, which is summarized in Table 2.2, research laboratories and institutions must establish ethics committees composed of five each of research scientists, laboratory technicians and lay persons. The committee must be

Table 2.2 Categories of experiments and their control in Sweden

Experiments requiring committee notification	*Experiments requiring committee approval*
1 No pain involved	4 Like 3 but with post-operative pain
2 Anaesthetized animals not revived after experimentation	5 Experiments expected to cause illness in anaesthetized animals
3 Anaesthetized animals revived without post-operative pain	6 Experiments using curare to cause immobility without anaesthesia

Adapted from Ross, M.W. (1978) 'The ethics of animal experimentation; control in practice', *Australian Psychologist*, *13*, 375–8.

notified of experiments in categories 1, 2 and 3 in Table 2.2 but, in addition, its *authorization* is also required before an experiment in categories 4 through 6 can be performed. In that case, a scientific or medical experiment on animals cannot, in principle, be conducted on the basis of scientific judgment alone. This is no guarantee of expert moral judgment, however, for qualifications of 'lay persons' in this respect may be no more than their qualifications as scientists. So who in the last analysis is the final moral judge? Lane-Petter (1976) provides an uncompromising answer.

> Not one of us can shrug off matters . . . such as this . . . we have a right and a duty to challenge the decisions that are made on our behalf. What the individual does not have the right to do is condemn, or refuse to listen to, those who have made moral or ethical judgments that differ from his own. (Lane-Petter, 1976, p. 121)

Whoever makes it, the ethical decision is frequently in the end a cost–benefit affair in which, by the nature of experimentation, the benefits are hypothetical while the costs are crystal clear. Were the results of an experiment to be fore-known the performance of the experiment would be a waste; and were the procedures of the experiment not disclosed the results of the experiment would be of no use. So resolution of the ethical dilemma necessarily involves balancing the unknown against the known, which is why there cannot be a categorization of permissible experiments after the manner of Table 2.2 based on the value of experimental results. It might seem that such a table could be constructed according to the practical potential of expected results (Seligman, 1975), but this is a retrogressive criterion because the power of scientific discoveries is to create practical possibilities, not to follow them. Had this criterion been adopted earlier, much applied medical and psychiatric research would never have begun, and its human beneficiaries would not be around to defend it.

In any event, as a factor in an ethical equation the dangers of the practicality criterion outweigh its gains. Experiments on the effects of high altitudes on the human body, on effects of freezing on warm-blooded creatures, on treatment for injury caused by mustard gas, on the efficacy of sulfonamides in treating gas gangrene war wounds, on factors affecting bone, muscle and nerve regeneration, on effects of drinking sea-water, and on possible vaccines for typhus, were all conducted on German concentration camp prisoners in World War II, and were all defended on the practical grounds of likely benefits to other humans – high-altitude flyers, shipwrecked sailors and wounded soldiers in the Nazi forces (Cohen, 1953). Those experiments were unethical in principle, regardless of their outcomes, practical or otherwise. As Pappworth (1967) points out, an experiment is ethical or not before it begins, not after it is over. 'Morality', he claims, 'rests on what is right in itself . . . not on justification by result, even though that may possibly benefit a great many others' (p. 185).

If we knew absolutely what is right in itself there would never be doubts about what is moral and ethical, but unfortunately, as Marcuse and Pear (1979) point out, that which a society calls right is that which it values most for its own survival. In the case of experiments that might do harm to animals, where the desire for knowledge and the importance of kindness to animals are especially valued, the practical question becomes that of formulating principles of humane experimentation, not of prejudging the practicability of likely outcomes or results. It is widely judged that an animal's life is less important than a human's and it is widely agreed that animals have the rights to human protection. It is the responsibility of scientists to preserve these rights to the limits of their abilities through the adoption of the most humane experimental procedures available at any time.

Humane experimental procedures

Ethical questions only arise in cases of imbalance of power and conflicts of interests. With laboratory animals, the imbalance of power is between the laboratory subject and the human experimenter, and the conflict of interest occurs when the animal is put in pain, discomfort or stress. These conditions are relatively common in experiments relevant to psychiatry, as in studies with aversive stimulation, infant–mother separation or social isolation rearing, because psychiatry specifically exists for individuals in distress. Such stressful conditions are not only found in laboratories, however. In farm management, lambs and calves are artificially reared, and painful electrical devices are used for fencing or for forcing cows to defecate in pre-arranged locations (Kiley-Worthington, 1977). For farmers, guidance for improvement of the welfare of their animals appears in connection with efficiency and productivity (MacDonald and Dawkins, 1981); guidance for experimenters is offered in connection with the humane arrangement of experiments, specifically with the *replacement* or *reduction* of animal subject populations and the *refinement* of experimental procedures (Russell and Burch, 1959).

Replacement involves either the use of less sentient for higher animals in experiments likely to cause distress, or the employment of alternatives to live animal experiments. Both of these raise the problem of generalizability of results that was discussed above, and each has its use in different particular areas. Replacement of animals in biomedical research is discussed by Smyth (1978), who notes that the principal alternatives – bioassay isotope tracing, chemical analysis by chromatography, and computer simulation – were all started as the purest of fundamental research, not in the remotest way related to the alleviation of human suffering. Some of these substitutes for animals may be useable for testing biological mechanisms of psychiatric drugs or for toxicity screening, but substitution of lower for higher animals in behavioural research only complicates the question of generalizing the results to humans. Although it is a laudable objective, it is unlikely that inanimate materials can soon substitute entirely for living animals as replacements for humans. Where research is on psychiatric problems of living, living organisms must necessarily be employed, and where the living organisms cannot be human for ethical or practical reasons, animal alternatives have to be found.

If animals cannot be replaced by inanimate materials in behavioural experiments to do with stress, at least the number of subjects employed can be kept to a minimum (Hume, 1957). The reduction of the number of subjects does not in itself make an unethical experiment ethical, for by Pappworth's (1967) principle of equality in human experimentation 'if it is unethical to submit many to a proposed experiment, it is equally unethical to expose only one person'. But reduction in subject numbers is a humane experimental procedure that can be combined with other refinements in procedures for animal experimentation.

Among such refinements, Russell and Burch (1959) recommend selective breeding of experimental animal subjects to minimize genetic contributions to variability in experimental data: This might serve some useful purposes, although it cannot substitute for improvements in experimental techniques and procedures. Psychology became an experimental science at a time when its capabilities of controlling a stimulus exceeded its capacity to control a response, so that precise manipulations of independent variables (stimuli) frequently produced imprecise dependent variable (response) data. The result of this was that experimental psychologists accepted experimental error (uncontrollable variance) as an inevitable characteristic of their discipline, and resorted to the employment of large subject populations to randomize individual variability and differentiate it from the true measures under investigation. Psychologists were forced into statistical control of error and opted to make descriptive and inferential statistics the cornerstones for the design of experiments.

This option is still widely employed in psychology, with the result that many experiments use large numbers of subjects just for the production of stable averaged data. However, this is no longer necessary because, with the precise control of behaviour obtainable with modern operant experimental techniques, error can be minimized by experimental instead of statistical refinement. In that case animals in experiments are employed not as groups for statistical control of inevitable experimental error, but as individuals, one by one, for the refinement and replication of experimental findings. When replications are made systematically by testing the generality of results under different conditions in successive experiments, fewer animals are necessary than when one experiment exactly replicates another. Thus by systematic replication the reliability and validity of experimental results can be assessed with the minimum number of experimental animals.

Just as experimental refinement can reduce the number of animal subjects in a particular study that are exposed to stress, so it can reduce the amount of stress to which each animal is exposed. In one common psychological procedure for assessing conditioned fear, for example, Davis and Wright (1979) found that a wide range of shock intensities was employed by different experimenters in arriving at much the same conclusions. On the other hand, with two similar procedures for studying shock avoidance learning by rats, Keehn (1967) found that the animals quickly learned to run in a wheel and avoid the majority of shocks, but failed to avoid the same level of mild intensity shock by pressing a bar. Thus if avoidance learning is a model for human anxiety neurosis (Levis,

1979) then it can be studied more humanely in rats with the running than with the bar-pressing response. Better still is the suggestion of Russell and Burch (1959) to look for a natural stressor for the animal concerned. In the screening of psychiatric drugs, for example, they illustrate how anxiety and fear-reducing agents can be assessed by observations of naturally opposed flight and courtship behaviours in fishes and birds, rather than by arbitrary laboratory stressors.

In practice, Ross (1981) has suggested the use of control procedures similar to those for Swedish experiments set out in Table 2.2 for behavioural research with higher animals. He recommends that wherever experiments with higher animals are proposed there should be a fifteen-member committee made up of five animal care technicians, five scientists and five other individuals. The committee should be readily available to make rapid decisions, and should be charged with two prime responsibilities: seeing that animals are not subjected to unnecessary harm or use; and seeing that research is designed to employ the fewest number of animals to the greatest advantage. A categorization of behavioural experiments in order of stressfulness to animal subjects proposed by Ross (1981) is summarised in Table 2.3, where studies in categories 4, 5, and 6 would automatically require full committee approval, while a research proposal falling in a lower

Table 2.3 Categories of experiments on higher animals

1	Painless experiments
2	Some distress is involved, including experimentation with non-optimum rearing conditions
3	Psychopharmacological trials and experiments employing aversive stimulation or non-laboratory reared animals
4	Aversive conditioning experiments and minimally painful experiments, CNS lesions or electrode placements which will cause minimal pain
5	Unavoidable pain, painful CNS lesions or stimulation, isolation studies
6	Long-term stress leading to animal neuroses and psychoses

Adapted from Ross, M.W. (1981) 'The ethics of experiments on higher animals', in Keehn, J.D. (ed.), *The Ethics of Psychological Research*, Oxford, Pergamon.

category could be authorized by fewer than all members of the committee.

As Ross (1981) concludes:

> The fact that such a research proposal has to be considered by such a committee frequently in and of itself has the advantage of sharpening the experimental design which is provided, and . . . the ethics of humane experimental technique may be administered without going to the extreme of either having a rigid set of rules or imposing control from outside the scientific community. (Ross, 1981, p. 59)

Given the size and significance of mental illness in human populations, it is essential for every effort to be made to bring the situation under control. Among these efforts is that which concentrates on the creation of animal models of the human illnesses. It is imperative that these creations do not themselves create more distress than the ones they are intended to eliminate, and it is encouraging to observe that experimental and administrative procedures are moving inexorably towards this end, fostered jointly though independently by the scientific and animal-loving communities.

Part II
Animal clinical pictures

3 Abnormal movements and convulsions

Stereotypies and bizarre postures

Definitions and examples

Stereotyped behaviours appear in autistic children, retarded children, psychotic adults and amphetamine addicts. In animals, stereotypies are induced by confinement, social and sensory deprivation, reinforcement schedules, and drugs. With humans, stereotypy pertains to symptomatology and diagnosis; with animals it relates primarily to origination.

Animals emit stereotyped behaviour in a variety of situations, of which pacing by zoo animals and rocking by a wild chimpanzee were described in Chapter 1. Other examples come from animals in laboratories and on farms. Kiley-Worthington (1977) observes that dogs, cats and horses develop stereotypies more frequently than cattle or deer, and lists variations of the kinds of situations in which stereotypies occur; restrictions on movement and sensory stimulation, confrontation with novelty, and conditioning. She defines a stereotypy as 'an aberrant behaviour repeated with monotonous regularity and fixed in all details'. However, not all stereotypies need be aberrant. So-called ethological stereotypies are fixed action patterns typical of a species. In the rat, for example, Barnett (1963) lists respiration, locomotion, ingestion, gnawing, hoarding, grooming, crawling under, fighting, coitus, parturition, nursing, retrieving and nest-building as stereotyped activities. These are not atypical activities of an aberrant individual but typical behaviour patterns of species members under specifiable conditions. Nevertheless, typical action patterns may occur atypically according to the form, frequency, consequence and mode of elicitation of the behaviour. Stereotypies often appear to be purposeless, but as Kiley-Worthington (1977) asserts, purpose is in the eye of the beholder and is best omitted as a defining characteristic of aberrant stereotypies.

Even the characteristics of stereotypies included in Kiley-Worthington's definition are not applicable to stereotypies of all kinds. Monotonous regularity is a particular characteristic of stereotypies induced by restrictions on territory or freedom of movement, but stereotypies that result from deprivation rearing are as likely to appear as non-repetitive stereotyped postures as they are to take the form of monotonous repetitive movements. Likewise the imperative that stereotypies are fixed in all their details is not universally applicable. Amphetamine-induced stereotypy, which is widely employed as a bridge to link animal responses to chronic amphetamine administration with secondary schizophrenic symptomatology, takes more the form of 'fragmented actions' than it does of monotonous movements. We must beware, then, of fitting all stereotypies into a single universal mould, and recognize their existence in several forms, not all of which are aberrant behaviours (e.g., ethological stereotypies). The particular categories of aberrant stereotypies I propose to describe are:

1 Cage stereotypies,
2 Deprivation stereotypies,
3 Aberrant postures,
4 Fragmented actions.

Originations in animals

Stereotyped behaviours have been differentiated according to form and according to origin. Differentiation by form is usually made with humans (see below), and the major categories are repetitive and non-repetitive movements. With animals, Berkson (1967) distinguishes between cage stereotypies and deprivation stereotypies according to the source of the stereotypy, and Robbins (1982) has introduced a category of fragmented actions to describe repetitive stereotypies induced by stimulant drugs.

CAGE STEREOTYPIES Sources of cage stereotypies may include food frustration (Kiley-Worthington, 1977), imminence of feeding time, or thwarting of natural flight reaction when danger signals appear (Meyer-Holzapfel, 1968). Pre-feeding stereotypies may be conditioned 'superstitious' responses (Skinner, 1948), or variants of sign-tracking behaviours described below. Keiper (1970) describes such a case of 'spot pecking' in caged canaries. This stereotypy is not affected by cage size, in contrast to a stereotyped 'route tracing' movement, which is.

Laboratory studies with monkeys show that cage stereotypies

are responsive to the size of the cages the animals are kept in (Draper and Bernstein, 1963) and also to the quality of the animal's environment (Berkson, Mason and Saxon, 1963).

Draper and Bernstein (1963) studied three male and nine female wild-born 3-year old rhesus monkeys in three outdoor cages of different sizes: small (3ft by 3 ft by 3 ft); medium (4 ft by 3 ft by 8 ft); and large (48 ft by 24 ft by 8 ft). Several categories of behaviour in the different cages are listed in Table 3.1, which shows that most stereotyped behaviour occurred in the small cage, that significantly less ($p < 0.01$) occurred in the medium cage, and that there was no stereotyped behaviour in the large cage at all.

Table 3.1 Most frequently observed stereotypical behaviours of twelve adolescent rhesus monkeys as function of size of holding cage

Small cage	*Medium cage*	*Large cage*
Stereotypies:		Self-directed activities
bouncing	hanging	grooming
pacing	sitting	self-clasping
twirling		self-biting
jumping		genitalia manipulation
somersaulting (backwards)		
Cage manipulation:		
shaking		
biting		

After Draper, W.A., and Bernstein, I.S. (1963) 'Stereotyped behavior and cage size', *Perceptual and Motor Skills*, 16, 231–4.

The stereotypies, which differed from animal to animal, took the forms of:

> rapid bouncing on the floor with all four feet, bouncing using only the front legs, predictable circular pacing, pacing with a head thrust at regular intervals, regular pacing and recoiling from one corner of the cage, rapid pacing developing into an exceedingly fast spin or twirl on the hind legs in the center of the cage, twirling holding onto the roof, backwards somersaults, unique awkward vertical jumping, and touching one leg to a particular place on the side of the cage as the animal travelled in a fixed pattern. (Draper and Bernstein, 1963).

Some of these behaviours, particularly awkward vertical jumping, are plainly adaptations to the tinyness of the small cage, and resemble the distorted intention movements of buntings kept in cages with perches too close to the ceiling (Hinde, 1962). The forms of other stereotypies, while responsive to cage size, were not specifically determined by the cage. For example, one female described by Draper and Bernstein showed continuous backward somersaults in the small cage, 'regular pacing that involved throwing up the forelegs and tossing back the head as if to begin the somersault', but without completing it, in the medium cage, and no sign of somersaulting in the large cage. Thus, somersaulting did not occur when it was easily possible, and occurred most when other movements were hampered.

In a set of four experiments, Berkson, Mason and Saxon (1963) observed variations in the stereotyped behaviour of four male and two female laboratory-reared adult chimpanzees in situations that differed in novelty, available space and opportunities for alternative activities. Five classes of behaviour (repetitive stereotypies, non-repetitive stereotypies, manipulation of the environment, self-manipulation, and locomotion) were recorded in an outdoor 39 ft by 57 ft home enclosure, a 69 in by 72 in by 85 in barred cage, and an 81 in by 79 in by 84 in enclosed wooden cubicle. The repetitive stereotypies observed were mostly rocking and swaying, but included also head nodding and shaking, and twirling. Non-repetitive stereotypies were abnormal limb postures, lip contortions, eye poking and thumb sucking. Self-manipulation occurred in the forms of scratching and rubbing. Most stereotypy, particularly of the repetitive kind, occurred in the cubicle, which was the smallest and the most isolated with respect to social and sensory communication, of the three environments. In a comparison of the cubicle and home enclosure environments, when manipulatable objects in the forms of a broom handle, a clothes line and a piece of burlap were and were not available, the availability of the objects was found to reduce non-repetitive stereotypies and self-manipulation by statistically significant amounts.

Davenport and Menzel (1963) compared stereotyped behaviours of sixteen chimpanzees raised from birth in various kinds of restricted environments at the Yerkes Laboratories with three wild-born chimpanzees brought to the laboratory at between about 4 and 7 months of age. The wild-born animals were kept together in a large open cage enriched with toys and exercise equipment, but the laboratory-born animals were housed individually and separately, except for two pairs whose individual cages were separated only by bars. Observations were made until the chimpanzees were over 40 months old, but almost no stereotyping was seen in the wild-born animals. By contrast, three classes of stereotypies were exhibited by the deprived chimpanzees: rhythmical rocking, swaying or body-pivoting; repetitive movements of head, hand or lips; and posturing in an awkward position.

Seven animals that were observed daily from birth to 21 months showed four major stereotypies that emerged at different ages: swaying, rocking, pivoting and thumb-sucking. The times of occurrence of these behaviours are indicated for each animal in Table 3.2.

Table 3.2 Month of onset and number of months up to 28 months of age in which each of seven chimpanzees raised individually in captivity was observed to display swaying, rocking, pivoting and thumbsucking stereotypies.

	Stereotypy							
	Sway		*Rock*		*Pivot*		*Suck*	
Animal	*Onset*	*Number*	*Onset*	*Number*	*Onset*	*Number*	*Onset*	*Number*
No. 200	8	16	8	5	3	15	2	7
No. 169	17	4	12	1	1	18	5	2
No. 171	11	10	8	8	4	17	2	15
No. 190	13	3	17	1	5	18	3	10
No. 173	9	3	5	19	4	25	2	4
No. 196	10	15	9	1	5	15	3	4
No. 188	8	19	12	6	6	10	2	12
Mean onset	11.0		10.0		4.0		2.7	
Mean number		10.0		7.2		17.0		7.7

Compiled from Davenport, R.K., and Menzel, E.W. (1963) 'Stereotyped behavior of the infant chimpanzee', *Archives of General Psychiatry*, *8*, 99–104.

Table 3.3 Frequencies of orality, disturbance and aggression stereotypies exhibited by eighty-four socially isolated rhesus monkeys under passive and stimulated conditions (see text for details)

Category of stereotypy	*Passive frequency*	*Active frequency*
Orality:		
Sucking orality:		
Toe sucking	1472	684
Other digit sucking	196	84
Self-sucking	141	36
Thumb sucking	50	72
Chewing orality:		
Chewing	3044	2795
Cage biting	440	52
Nail biting	166	10
Other orality:		
Self-licking	395	24
Cage licking	238	4
Finger in mouth	12	2
Mouth rubbing	3	2
Tongue pulling	1	0
Disturbance		
Vocalization	553	1879
Grimacing	50	1946
Head lowering and leg clasping	464	524
Rocking	358	708
Self-clutching	96	528
Convulsive jerking	48	248
Aggression		
Externally directed:		
Threat	162	2507
Cage shaking	146	382
Self-directed:		
Displaced threat	220	1188
Self-biting	294	1092
Teeth grinding	40	472
Head slapping	13	156

Condensed from Cross, H.A., and Harlow, H.F. (1965) 'Prolonged and progressive effects of partial isolation on the behavior of macaque monkeys', *Journal of Experimental Research on Personality*, *1*, 39–49.

DEPRIVATION STEREOTYPIES Stereotyped activities also occur in laboratory-reared monkeys that are socially isolated either as a matter of routine care and maintenance or for specific experimental purposes. Table 3.3 lists stereotypies observed by Cross and Harlow (1965) under passive and stimulated conditions of observation. They studied *orality*, *disturbance*, and *aggression* categories of stereotypies in eighty-four socially-isolated rhesus macaques ranging from 1 to 7 years of age. The table shows the number of times that each listed stereotypy occurred among all of the animals during ninety half-minute observations per animal over a period of three weeks. In the passive condition, observations were made unobtrusively; in the stimulated condition a threat stimulus in the form of a large black glove, usually used for handling the animals, was slowly moved across each animal's cage while it was under observation. Behaviours that differed significantly according to sex and age are marked in Table 3.4.

Among the abnormal behaviours observed by Cross and Harlow (1965) were stereotyped pacing and back-flipping, self-clasping, finger-sucking and body-rocking, which are typical stereotypies induced by captivity and confinement. Other observations, of withdrawn-activity, self-biting, hair-pulling, self-mutilation and apparent dissociation, are more characteristic of social deprivation.

As well as with monkeys and chimpanzees, infant social and sensory deprivation has been studied extensively in rats, mice and dogs (Beach and Jaynes, 1954). Mostly these studies have been statistical probes for critical periods in the development of social and problem-solving capabilities (Scott, 1962), but occasional reports of specific aberrant behaviours are available. In one, Melzack and Scott (1957) described how 9-month-old Scottish terrier puppies raised for seven months in almost total sensory isolation repeatedly burned themselves on lighted matches held before their noses. The dogs 'moved their noses into the flame as soon as it was presented, after which the head or whole body jerked away . . . but then they came right back to their original position and hovered excitably near the flame.' In another study, Thompson, Melzack and Scott (1956) report 'whirling behavior' in eight of eleven other puppies raised under similar conditions. The whirling stereotypy consisted of 'very rapid, jerky running in a tight circle' accompanied by yelping, barking, snarling and tail-biting sometimes lasting as long as ten minutes.

ABERRANT POSTURES Aberrant bizarre postures such as fixed staring, rigidity, awkward mobility and huddling do not qualify as stereotypies in the monotonous repetitive sense, but they are not uncommon in human clinical types in which repetitive stereotypies

Table 3.4 Stereotypy categories showing statistically significant age and sex differences by eighty-four socially isolated rhesus monkeys

	Greater in:		Greater in:	
Category	*Older*	*Younger*	*Males*	*Females*
Orality				
Sucking orality:		X		
Toe sucking		X		
Chewing orality:				
Chewing	X			
Cage biting		X		
Nail biting				
Other orality:				
Self-licking disturbance				
Vocalizing		X		X
Grimacing		X		X
Head lowering and leg clasping		X		
Rocking		X		
Self-clutching		X		
Convulsive jerking		X		
Over-all disturbance aggression		X		
Externally directed:				
Threat	X		X	
Self-directed:				
Displaced threat	X		X	
Self-biting	X		X	
Teeth grinding			X	
Head slapping			X	
Over-all self-aggression	X			

Condensed from Cross, H.A., and Harlow, H.F. (1965) 'Prolonged and progressive effects of partial isolation on the behavior of macaque monkeys', *Journal of Experimental Research on Personality*, *1*, 39–49.

also appear. In an animal, Merlin (p. 13) is an illustration of the co-existence of a bizarre posture (hanging upside-down) and a repetitive stereotypy (rocking).

The most common origination of such aberrations in laboratory animals is social deprivation or inadequate rearing, and Mason (1968) lists the first of four characteristics of a primate deprivation syndrome: abnormal postures and movements; motivational disturbances; poor motor integration; and deficiencies in social communication. A vivid illustration of bizarre posturing in a monkey raised in isolation is a case recounted by Mitchell (1970). The animal could also be exhibiting a delusion, a hallucination or dissociation.

> One . . . male slowly moved his right arm toward his head while in a rigid seated pose and, upon seeing his own approaching hand, suddenly appeared startled by it. His eyes slowly widened and he would at time fear grimace toward, threaten, or even bite the hand. . . . If he did not look directly at the hand or did not bite it, 'it' would continue to move toward him. . . . As 'it' approached him, his eyes became wider and wider until the hand was clasping his face. There he would sit for a second or two, with saucer-sized eyes staring in terror between clutching fingers. (Mitchell, 1970, p. 228)

Social or sensory isolation is not the only condition that occasions bizarre posturing in animals. The co-occurrences of aberrant postures and repetitive stereotypies is reported by Davenport and Menzel (1963) in non-isolate laboratory-reared chimpanzees. These authors categorise rhythmic *whole-body movements* (rocking, swaying, somersaulting, chest pounding), *part-body movements* (head banging, thumb and toe sucking, hand clasping) and *posturing* (head rolling, staring at the hand before the eyes) all as stereotypies. They report that these related not only to rearing but also to developmental status as summarized above in Table 3.2. For Davenport and Menzel (1963), as for Meyer-Holzapfel (1968) in connection with zoo animals, space restriction is the most striking contribution to stereotypies and bizarre posturing in animals. In such cases, as with animals on farms (Kiley-Worthington, 1977), scientific discoveries are put to use in improving the care and welfare of animals as well as in providing clues for improving the lot of humans.

FRAGMENTED ACTIONS Robbins (1982) coined the term fragmented actions to explain a theory that he and Lyon (Lyon and Robbins, 1975) had proposed to account for the action of amphetamine-like drugs, and to explain the correspondence between certain behavioural results of abuse of these drugs and schizophrenia.

The theory stems from three observations. The first is that in humans amphetamine overdose produces a clinical picture that can be confused with paranoid schizophrenia (Connell, 1958). The second is that stereotypy is a salient secondary characteristic of schizophrenia (Bleuler, 1950; see below). And the third is that animals given high doses of amphetamine and similar drugs exhibit perseverative stereotyped responses. These responses are not the characteristic pacings and rockings of caged or socially deprived animals but stereotyped intrusions into behaviour sequences that, if completed, would lead to a satisfactory end. That is, animals trained to perform a sequence A–B–C to secure food under ordinary circumstances might, after continuous amphetamine injections, perseverate in early portions of the sequence and emit AAAABAAABBABC instead. The stereotypies need not be productive parts of the behaviour chain but species-specific drug-induced competing responses, like sniffing and headshaking in the rat, that stop the chain part way and start it over from the beginning. At low doses amphetamine behaves as a stimulant by elevating gross motor activity in all parts of the sequence, but as drug use and dose increase 'the increasing acceleration of behavioral initiation results in an increasing repetition within a decreasing number of response categories' (Lyon and Nielson, 1979, p. 110). This is behavioural fragmentation.

The response categories into which amphetamine-induced stereotypies fall are dose-dependent. As Randrup and Munkvad (1975) report them for the rat:

> 1 mg/kg of d-amphetamine given subcutaneously to rats thus produces selective stimulation of sniffing, locomotion and rearing while there is little grooming activity. . . . Gradual increase of the amphetamine dose in the interval 1–10 mg/kg leads first to a further decrease and eventually to disappearance of grooming activity; then locomotion and rearing are also decreased and finally disappear so that only sniffing remains. At 10 mg/kg the maximal stereotypy is reached. Sniffing, often accompanied by licking and biting, is performed continuously and usually covers only a small area at or near the bottom of the cage. (Randrup and Munkvad, 1975, p. 759)

This account of amphetamine-induced stereotypies points to dose-dependent selective *suppression* of ethological stereotypes in the rat (Barnett, 1963, see p. 49), 'so that only sniffing remains'. This is the stereotypy to which Robbins (1982) attributes maximum behavioural fragmentation in that animal. Amphetamine-induced stereotypies reported for other species are looking from side to side in cats, circling or running back and forth in dogs, twittering and posturing in chicks, biting and chewing in the tortoise and the lizard, and pecking in several species of birds (Randrup and Munkvad, 1975).

Going from rodents to humans, stimulant-drug stereotypies take more and more complex forms. Randrup and Munkvad (1975) classify these forms as ethological, operant, emotional and social stereotypies, with 'hung up' repetitious human thoughts and acts as the most complex. A study with forty-six cats by Ellinwood, Sudilovsky and Nelson (1972) assessed movements, postures and 'attitudes to the environment' as a function of methamphetamine exposure and dose. Concerning movements and postures, they recorded head–neck, shoulder–foreleg, hip–hindleg, tail and trunk bodily regions, and assessed overall coordination and synchrony between the regions. They found that over eleven days of chronic methamphetamine administration, movements in the respective regions became less coordinated and more dysynchronous. The dysynchronies in movement, where, for example, frozen hindlegs occurred concurrently with hyperactivity in head and forelegs, eventually led to a disjunctive posture wherein 'the active forelegs at times would back up against the relatively resistant hind legs to produce the hunch or camel-back phenomenon'. Beyond that, Ellinwood *et al.* continue: 'it appears as if the cat had forgotten where a leg is positioned . . . [it] would remain in an awkward disjunctive position while the cat goes about other activities.'

The 'attitudes to the environment' category included awareness, indifference, normal and abnormal interest, normal and abnormal investigative behaviour, reactivity, and normal and abnormal focus (apprehensively 'hooked'), all subjectively assessed on the basis of an animal's bodily posture in relation to its surroundings. In this category Ellinwood *et al.* record that methamphetamine induced a compulsive investigative fixated interest by the animals to a constricted sector of the environment.

All in all, Ellinwood *et al.* (1972) argued that their amphetamine-intoxicated cats show many resemblances to human catatonics, and conclude that:

> Following chronic amphetamine intoxication, components of behavior became relatively fixed over time and showed a loss of cohesive flow among different initiatives with their relative priorities. In addition . . . there appeared to be islands of separate organization, each establishing its own autonomy or anarchy without integration into the larger behavioral symphony. (Ellinwood *et al.*, 1972, pp. 227–8)

The fragmented actions that accompany amphetamine intoxication in animals and man are offered by many investigators as evidence for an amphetamine model of human schizophrenia founded on the distribution of catecholamine neurotransmitters in the brain. Comparative evidence in support of the model from acute, progressive and residual effects of high-dose administration of stimulant drugs with rats, cats, monkeys and humans is summarized in Table 3.5 (Ellinwood and Kilbey, 1977). There are, however, critics of the model (Kokkinidis and Anisman, 1980).

Symptomatology in humans

Four classes of humans exhibit stereotyped behaviours: retardates, autistic children, schizophrenics, and amphetamine addicts. In these cases the origination of the disorder is a matter of conjecture, and the behaviours are employed as criteria for differential diagnosis rather than as indicators of deficiencies in conditions of living. In these humans, aberrations that in animals are attributed to short- or long-term environmental circumstances are more usually attributed to personality defect, organic defect, or mental illness. Stereotypies in humans are more complex and varied than those that appear in animals.

RETARDATES Berkson and his associates (cf. Berkson, 1967) report extensive studies of stereotyped movements in mentally defective humans. In children and adults they identify *repetitive movements* of body (swaying, rocking, twirling), head (shaking, nodding, banging), face (tics, grimaces), and hands (fiddling), and *non-repetitive postures* and self-manipulation (rubbing, scratching, poking, biting and picking). With respect to repetitive movements, some individuals appear mostly as rockers and others as fiddlers. These behaviours also appear in normal children but they do not persist. With the retardates, general arousal level and opportunities for competing behaviours affect the prevalence of stereotypies although, as Berkson (1967) concludes, the effects are ephemeral.

Stereotypies in retardates have been studied extensively for therapeutic and institutional ward management purposes, and have been shown to respond to sensory and social environments. Luiselli (1975), for instance, reports on an institutionalized 14-year-old boy with a 50–60 IQ range who engaged in frequent rhythmic rocking wherever he happened to be sitting. The boy's rocking was systematically observed and counted on a number of daily occasions (baseline) and was then subjected to a number of treatment modes. In the first, the boy received social praise, popcorn and candy when he was sitting still without rocking, while rocking was ignored; in the second, the attendant walked off and

Table 3.5 Characteristic acute, chronic and residual effects of high-dose administration of stimulants to humans (clinical reports) and animals (experimental reports).

	Acute effects		
	Stereotypy	*Movement*	*Attitude*
Human	Repetitive movements Compulsive acts		Suspicious
Monkey	Visual scanning		Investigatory
Cat	Head movements	Dysjunctive	
Rat	Sniffing Licking Gnawing	Restricted	
	Chronic effects		
	Stereotypy	*Movement*	*Attitude*
Human	Oral dyskinesias Facial tics	Choreiform movements Catatonic immobility	Paranoia Delusions of parasititis
Monkey	Oral dyskinesias Constricted, bizarre Self-grooming	Dystonic posture	Reactive Socially inadequate
Cat	Paw shaking Head shaking	Akathesia Dystonic posture Ataxia	Hyper-reactive
Rat	Jerky Increased intensity	Backward walking Jumping	Reactive
	Residual effects		
Human	Low dose thresholds for induction of psychosis or dyskinesias.		
Monkey	One or two doses reinstate chronic end-state behaviours.		
Cat	Low dose reinstatement, conditioned behaviour to setting.		
Rat	Persistent augmented response to usual dose.		

Adapted from Ellinwood, E.H., and Kilbey, M.M. (1977) 'Chronic stimulant intoxication models of psychosis', in I. Hanin and E. Usdin (eds), *Animal Models in Psychiatry and Neurology*, New York, Pergamon Press.

ignored the boy at the onset of rocking and returned with praise and candy when the rocking stopped. In the third treatment condition proper sitting continued to be socially reinforced, but the boy was sent to stand in a corner for 3-minute time-out periods whenever he rocked; that is, reinforcement and punishment procedures were combined. The maximum 'therapeutic' effect occurred with the last procedure, and suppression of rocking was maintained in subsequent baseline and time-out plus reinforcement phases.

As with animal stereotypies, stereotypies of retarded humans are amenable to environmental manipulation, although whereas with animals the manipulations are for analytical and experimental purposes, for humans they are for the purposes of treatment and institutional management.

AUTISTIC CHILDREN Early infantile autism is a syndrome identified by Kanner (1943) and distinguished from childhood schizophrenia on the basis of early age of onset, resistance to change in the environment, and intellectual-type personalities in the parents. Others have proposed different criteria of autism (see Chapter 8), and the syndrome is not reliably diagnosed. However, bizarre gestures and stereotyped repetitive movements are uncontested characteristics of autistic children. Epstein, Doke, Sajwaj, Sorrell and Rimmer (1974) describe vacant staring and grimaces, rocking, swaying, and object twirling as characteristics of autistic children, and classify autistic stereotypies as:

1 inappropriate foot movements (jumping, hopping, running);
2 inappropriate hand movements (pounding, spinning, flapping, rubbing); and
3 inappropriate vocalizations (echolalia, talking to self, mumbling).

Interpretations of these behaviours usually stress self-stimulation and self-reinforcement. They are generally unresponsive to environmental control, but some stereotypies, such as head-banging, can be suppressed by contingent punishment (Lovaas, Schaeffer and Simmons, 1965), and others may be maintained by social reinforcement (cf. Lucy, p. 50). A little 4-year-old girl, for instance, whom I shall call Dorothy Rocker, was hospitalized for infantile autism on account of the usual criteria (see Chapter 8). She performed an occasional repetitive stereotypy of crouching on all fours and making pelvic thrusts of an unmistakable sexual nature. The behaviour was selected for analysis, and for brief daily periods she was placed in a room with some toys and observed through a one-way mirror. Her stereotypy began soon after she entered the room, and it was duly counted, but it soon became evident that, as in the case of the chimpanzee, Lucy, the stereotypy was situation-dependent: Dorothy constantly looked toward the mirror while she was rocking, and as soon as the door opened at the end of the session she stopped rocking and ran into the hall. In such events it seems preferable to analyse autistic stereotypies in humans in the way they are analysed in lower animals – case by case on an individual basis rather than in terms of a general characteristic of a clinical syndrome of autism.

AMPHETAMINE ABUSERS Soon after World War II reports from Japan drew attention to instances of 'incomprehensible, odd and very unnatural movements' that are 'constantly, identically and energetically repeated' by amphetamine abusers (Randrup and Munkvad, 1967). In similar vein, Randrup and Munkvad describe cases in Sweden where abuse of an amphetamine-like compound, phenmetrazine, produced 'compulsive or automatic continuation for hours of one aimless activity, such as sorting objects in a handbag, manipulating the interiors of a watch, polishing finger-nails to the point that sores are produced, etc.'

Other reported illustrations of amphetamine-induced stereotypies in humans are teeth-grinding, chewing, lip movements and 'hung-up' repetition of simple thoughts. Similar symptoms appear in schizophrenic patients (see below), which has caused misdiagnosis of amphetamine intoxication as paranoia (Connell, 1958). From this observation has developed the amphetamine model of schizophrenia.

SCHIZOPHRENICS Many of the aberrant behaviours described above for autistic children also appear in clinical accounts of schizophrenic adults. Tics, pacing, repetition of words and movements are characteristics of schizophrenics, as also are stereotyped thinking and echolalia. Aberrant postures like adoption of a foetal position or imitation of a Christ-like posture with arms outstretched as if on a cross are textbook illustrations of schizophrenic behaviours, and Bleuler (1950) includes stereotyped postures and repetitive movements as among the striking secondary characteristics of schizophrenia.

The correspondence between amphetamine-originated stereotypies and symptomatogical stereotypies in psychosis is the basis of the dopamine theory of schizophrenia reviewed by Silverstone and Turner (1982). This theory addresses the problem of schizophrenia indirectly by noting first the equivalence (in part) of symptomatologies of unknown origin (clinical stereotypies) and of known origin (amphetamine stereotypies), and second, the role of dopamine in the action of amphetamine in the brain and central nervous system. The second is the step expected to expose the biochemical bases of human psychoses with animal models such as those described in Chapter 7.

Abnormal fixations and seizures

Compulsion and insoluble problems

IN IMPOSSIBLE DISCRIMINATIONS Although well known in farm animals for ages, reports of stereotypies in laboratory animals are relatively recent, and consideration of these animal behaviours as models of human psychopathology is an even later occurrence. In the case of so-called experimental neurosis, however, while farm animals were among the early subjects of investigation (Liddell, 1944), psychological interest quickly turned to the laboratory rat (Finger, 1944). At the time, behavioural disorders were primarily attributed to conflict, and conflict became the focus of laboratory studies with the rat. Such studies did not generate an acceptable animal model of human neurosis, but they nevertheless did generate behavioural abnormalities in the rat worthy of independent investigation.

So-called neurotic disablement caused by an insoluble problem was discovered inadvertently by N.R.F. Maier (1949) in the course of investigations of discrimination learning by rats with an apparatus devised by K.S. Lashley. In this apparatus, subjects must jump from a platform through the one of two doors that hides a food reward. The incorrect door is locked and mistakes give the animals a bumped nose and a short fall into a net. When the doors are discriminable rats jump promptly to whichever side is correct on a training trial, but when discrimination is impossible they show reluctance to jump and, when pushed fixate their responses on a single side. These responses are called 'abnormal fixations', although under the circumstances they are the most efficient reactions the rats can make. When jumps to either side are reinforced half the time at random, the rate of reinforcement is maximized if all the jumps are made to one of the sides. In that way half the jumps are reinforced for certain, a fraction that is beatable only by accident with random jumps. The reinforcement schedule for stereotyped 'fixated' left or right jumping is *variable ratio two* meaning that on average every second response secures reinforcement.

Impossible discriminations were originally conceived as insoluble problems, not as reinforcement schedules, and the rats' fixations were attributed to frustration-induced disablement. Some animals seemed to be upset by the experience, but, Maier (1949) says, 'the fixated group develops some kind of adjustment to the test situation and is therefore able to prevent emotional tensions'. With soluble after insoluble discriminations, fixated rats continue to make stereotyped position choices, but they jump differently to each of the discriminable stimuli: 'When the rat jumps to the rewarding window, it does it head-on, and when it jumps to the punishing window, its long axis becomes parallel to the window so as to avoid a painful blow' (Liberson, 1967). The fixated rat, like the neurotic patient, seems to make its responses against its will, which testifies to the power of the varible ratio schedule of positive reinforcement as much as to conflict-induced frustration.

IN REINFORCEMENT SCHEDULES A monkey trained by Findley and Brady (1965) showed even more compulsive-looking behaviour than that exhibited by Maier's rats. Hour upon hour the monkey sat pushing a button, a compulsion with no seeming effect. However occasional responses turned on a light-bulb and after 40,000 presses the monkey received a meal. The behaviour did not develop by accident but was carefully trained with schedules of fixed ratio reinforcement that got longer and longer day by day. The light served as a conditioned reinforcer and originally signalled a meal, but with training, fewer and fewer light onsets were followed by feeding till finally the ratio was 1 in 10. By the end, it took 4,000 responses to produce a light-flash, and after every 10th light the monkey was fed. A diagnosis of compulsive pushomania is an appealing evocative description of the monkey's *behaviour*, but it does not categorize the *monkey*, nor does it indicate how the 'pusho' is caused. Even a patient observer, seeing the monkey only after button-pushing developed, might never discern the origin of the animal's compulsion or how it was being maintained by a *fixed ratio 40,000* schedule of reinforcement. He could hardly be unimpressed by the amount of energy expended by the animal for so very little reason.

IN A SOCIAL SITUATION Calhoun (1967) describes a compulsive behaviour that also has an element of sadism, although its genesis is perfectly normal. The experiment was one of social cooperation among rats. In one group (COOP) two rats were required to cooperate in operating a water dispenser before either rat could drink; in another group (DISOP) no rat could drink if more than one approached the dispenser at a time. Calhoun describes one rat from the DISOP group that kept climbing a barrier separating the pens of the two groups.

> He would enter and approach the lever [of the COOP group water dispenser, and] one of the COOP rats would come over and enter the other side. To this invading DISOP male, his COOP companion's behavior was all wrong. He would immediately back out, grasp the 'offending' COOP rat by the tail or hind feet and pull him out. . . . During the following weeks he macerated the tails and hind feet of all the COOP rats. Most lost all their toes. Seven died from these wounds. And yet the invading male was never attacked. To the COOP rats this invading DISOP male was always behaving correctly . . . and their ethical standards dictated that they came to his rescue. (Calhoun, 1967, p. 20).

All these illustrations of compulsive behaviours in animals have parallels in human counter-productive compulsive actions. The animal cases may not have the full quality of the human 'symptom' but they do have the counter-advantage of illuminating originating situations: random reinforcement, ratio reinforcement, and a change in the social reinforcing context. These kinds of situations are not uncommon in human experience and they undoubtedly contribute to human compulsive behaviour. Such behaviours are not necessarily 'neurotic', however; that depends on the consequences of the act on society, and the reaction of society to the actor.

Superstition and reinforcement schedules

Experimental distress or disablement can be operantly conditioned by suitable reinforcement schedules. These are rules about contingencies between specified responses and environmental events like delivery of food or electric shock. Operant conditioning with a variety of schedules of reinforcement is described by Ferster and Skinner (1957). The formal procedure involves individual animals working in isolated 'Skinner boxes' with reinforcers programmed for specified responses on scheduled occasions, such as every other response or the first response in every minute. In addition to those mentioned in the section above, several schedules of reinforcement have disabled animals or caused them to behave strangely or with maladaptive emotion. The following are the principal examples:

1 *Fixed time*, in which reinforcers are delivered at short predetermined intervals independently of an animal's behaviour.

2 *Fixed or variable ratio*, in which reinforcers are delivered after an animal makes a regular or irregular number of responses.
3 *Fixed or variable interval*, in which reinforcers are scheduled for the first response that is emitted after regular or irregular time intervals.
4 *Multiple schedules*, in which two or more simple schedules alternate such that reinforcement density differs in the various components, or such that punishment and positive reinforcement are mixed.
5 *Conditioned emotional response schedules*, in which periodic predictable or unpredictable electric shocks are imposed upon other reinforcement schedules, usually variable interval reinforcement with food.
6 *Escape schedules*, in which responses terminate aversive events like electric shocks, loud noises or bright lights.
7 *Avoidance schedules*, in which animals can avoid or postpone aversive events that would otherwise recur with or without warning.

These schedules may maintain abnormal behaviour either by indirect induction of emotional behaviours, (schedule-induced behaviours, see Chapter 4), by deliberate reinforcement of specific unusual responses, or by accidental response-reinforcement contingencies (superstitious behaviours).

Laboratory-conditioned operants are usually prosaic efficient responses, like bar presses by rats and monkeys or key-pecks by pigeons, but Skinner (1948) once conditioned bizarre helpless-looking behaviour in pigeons by delivering reinforcers at short regular intervals regardless of what the birds were doing. The schedule is one of *fixed-time reinforcement* and Skinner describes the ensuing behaviour as superstitious. He writes;

> One bird was conditioned to turn counter-clockwise about the cage, making two or three turns between reinforcements. Another repeatedly thrust its head into one of the upper corners of the cage. A third developed a 'tossing' response, as if placing its head beneath an invisible bar and lifting it repeatedly. Two birds developed a pendulum motion of the head and body, in which the head was extended forward and swung from right to left with a sharp movement followed by a somewhat slower return. The body generally followed the movement and a few steps might be taken when it was extensive. Another bird was conditioned to make incomplete pecking or brushing movements directed toward but not touching the floor. None of these responses appeared in any noticeable strength during adaptation to the cage or until the

food hopper was periodically presented. (Skinner, 1948, p. 168)

Although the experimental subjects seem to behave strangely in comparison to other birds, their oddities are not neurotic oddities in the usual sense. They are specific oddities that Skinner claims are learned, and the origination of the oddities in the experiment seems clear. However, the superstitions may not look as weird to other pigeons as they do to human experimenters, because hungry pigeons waiting for food often do the things that Skinner observed.

Audiogenic seizures

IN BEHAVIOUR-GENETIC ANALYSIS Audiogenic seizures in rats and mice have been elicited by hissing airblasts, buzzers, high-pitched tones, bells, and jangling keys. Finger (1944) characterizes the audiogenic seizure as a three-part phenomenon in the rat; a fore-period followed first by an active and then by a passive phase. In the fore-period, after an initial startle to the noise, the rat exhibits crouching, burrowing, and a state of heightened sensitivity to noise. Sometimes, especially when a seizure does not occur, the rat performs 'substitute activities' of nose rubbing, ear scratching, teeth chattering, grooming and yawning; otherwise there are pivoting movements of the head and body, jerkiness and short quick runs that lead to a full-blown active seizure. In this active phase there occurs an explosive bout of running after which the animal falls down and emits a series of rapid clonic twitches. There may also be spasmodic bursts of running and jumping along with ejaculation, defecation and squealing. Occasionally with young animals this phase ends in death, otherwise the animal enters the third, passive phase. In this phase the animal remains comatose with no spontaneous activity and depression of normal reflexes for a period from two to ten minutes. After this, Finger and Schlosberg (1941) report, subnormal activity may persist for another twelve hours.

However, not all rats exhibit the phenomenon, and a particular rat may not convulse on every exposure to the stimulating noise. With a high-pitched whistle as the eliciting stimulus, Auer and Smith (1940) reported repeated convulsions in 30 per cent of a group of over 400 rats. Some quantitative features of typical convulsions of ten randomly selected animals are shown in Table 3.6. The table summarizes latencies and durations of convulsive running, and durations of jumping and rigidity during the periods of stimulus onset shown in the right-hand column.

Audiogenic seizures have served animal experimentalists as a

*Table 3.6 Quantitative characteristics of the different phases of the convulsive pattern produced in rats by a high-pitched stimulus whistle. Data are from a random selection of ten rats from seventy-five convulsive animals. Stimulation was usually terminated during the convulsive jumping stage, so durations of this stage are arbitrary (*time given in seconds*)*

Subject	*Latency of convulsive running*	*Duration of convulsive running*	*Duration of rigid state*	*Duration of convulsive jumping*	*Duration of stimulus*
1	44	15	12	175	246
2	50	40	43	147	278
3	46	14	29	70	289
4	45	7	25	125	220
5	5	43	5	168	213
6	7	29	8	155	239
7	35	23	33	100	213
8	5	18	21	160	211
9	8	11	20	155	400
10	10	10	17	330	373

From Auer, E.T., and Smith, K.U. (1940) 'Characteristics of epileptoid convulsive reactions produced in rats by auditory stimulation', *Journal of Comparative Psychology*, 30, 255–9.

convenient phenotype for the study of behavioural genetics (Fuller, 1979), and also as models for epilepsy and the alcohol withdrawal syndrome in man. Concerning behaviour-genetic analysis, Fuller (1979) distinguishes between two kinds of phenotypes: *somatophenes*, to do with physical structures; and *psychophenes*, to do with behavioural processes. The psychophenes he divides into an *ostensible* class of observed behaviours and an *inferred* class of generalized tendencies or traits. The audiogenic seizure is a member of the ostensible psychophene class inasmuch as it is an observable, countable behaviour.

Particular genotypes need not have identical phenotypes. Fuller points out that inbred strains of mice, for instance, of identical genotype may have the same pigmentation but differ in size, growth and ability to learn. The colouration he calls phenostable, the others, phenolabile. Phenolabile traits or characteristics have a reaction norm and a reaction range which are dependent on the nurturing environments of the group and of the individual. Thus the location of a phenotype on the dimension of phenostability–phenolability will determine the degree to which that phenotype is amenable to genetic and environmental control.

Detailed behaviour–genetic analysis of seizures with rats have yet to be reported, but the general role of heredity in audiogenic seizures has been known almost from the discovery of the phenomenon in that species. Thus Maier and Glaser (1940) found 74 per cent, 52 per cent and 0 per cent of 3-month-old offsprings of seizure susceptible–susceptible, susceptible–normal, and normal–normal crosses, respectively, to be classifiable as seizure-susceptible. Likewise there are strain differences in susceptibility to seizures among mice, where the susceptibility is higher among DBA/2J strain members than among members of the BALB/CJ and C57BL/6J strains (Fuller, 1979).

REFLEX EPILEPSY Extensive studies of audiogenic seizures in rats selectively bred for susceptibility to seizures are reported by Krushinskii (1962), who coined the term 'reflex epilepsy' in 1949. Krushinskii and his colleagues conceive of audiogenic seizures according to the Pavlovian conception of higher nervous activity as interactions between excitatory and inhibitory processes (see Chapter 5), and they employed, among other things, inhibitory and excitatory drugs to manipulate the inhibitory–excitatory cortical balance in their subjects. They used a loud-pitched noise (14–16 kHzs; 70–112 db) to engender convulsions, which were scored on a five-point scale from zero for no reaction to 4.0 for a complete tonic convulsion within one and a half to two minutes after noise onset.

With sodium bromide to strengthen inhibitory cerebral processes, Krushinskii reports a weakening of convulsions with increasing bromide dose. Conversely, with caffeine and strychnine, employed to increase excitatory processes, Krushinskii reports a statistically significant shortening of latencies of convulsions in comparison to normal non-drug measures. On the basis of such findings, Krushinskii recommends the audiogenic seizure response of rats as an experimental preparation for the rapid evaluation of drugs newly synthesized for the treatment of epilepsy.

IN ALCOHOL WITHDRAWAL The alcohol withdrawal syndrome is described in Chapter 6, where correspondences in the syndromes of rats, dogs, primates and humans are detailed. In animals, the stages and complexities of the syndrome are less definable than they are in humans, but they all include tremors, hyperexcitability and convulsions in the last degree. In alcohol-withdrawn rats, audiogenic convulsions have been elicited by the noise of jangling keys. Falk, Samson and Winger (1972) offer this as evidence of physical dependence on alcohol by rats made polydipsic by the spaced-feeding technique (see Chapter 4, p. 81). Their animals were fed small food pellets at 2-minute intervals for 1 hour periods every 3 hours, day and night, with only alcohol to drink. After about 2 months, four animals were subjected to tests for audiogenic seizures 3 or 4 hours after alcohol was withdrawn. Falk *et al.* (1972) give this description of the result.

> A shaking of keys near the top of the cage for 1 to 2 seconds resulted in a tonic–clonic convulsion in rat No. 8. For the next hour, tremors, spasticity, and clonic head movements occurred, and finally, a second seizure ended in death. When keys were shaken (2 to 5 seconds) for the first time after 9½ hours of withdrawal, a clonic running episode was produced in rat No. 2, followed shortly by death from a tonic–clonic seizure. Rat No. 7 showed all the preconvulsive symptoms, but keys shaken (up to 20 seconds) after 15 hours of withdrawal had no effect. (Falk *et al.*, 1972, p. 813)

The fourth rat was spared the key-shaking test, which had no effect on normal rats of the same strain that had not been subjected to the alcohol intake training and withdrawal routine beforehand.

The kindling effect

AS A MODEL OF EPILEPSY The kindling effect refers to the gradual appearance of stereotypies (behavioural automata) and convulsions in rats, monkeys and cats after a number of brief low-intensity electrical stimulations of the amygdala or other regions of the brain. Stimulations can occur from a few minutes to several days apart, and the effect generally appears in three successive stages (Gaito, 1979). Stage I is the appearance of apparently normal exploratory behaviour (ethological stereotypies) during the first few stimulations; Stage II is the emergence of preconvulsive stereotypies of chewing, salivation and eye closure; and Stage III is the final convulsive stage in which the rat rears up and begins to clonically convulse with the forelegs. The convulsions do not stop immediately the stimulating current is terminated, and the effect of kindling may appear in spontaneous convulsions in rats subjected to many stimulating trials spread over four to five months. Thus the effect is relatively permanent, a conclusion supported by the fact that after a period of up to several months without stimulation, re-kindled convulsions occur after only a few trials, or even on the first.

Kindled seizures in animals and epileptic seizures in humans bear some resemblance to each other, although that need not make them analogous. Nevertheless, the kindling procedure is capable of stimulating an epileptic-like condition, and Gaito (1976) describes two ways in which the kindling effect might contribute to solving the problem of epilepsy: first by revealing brain chemical and structural features involved in kindled convulsions; and second by uncovering possible therapeutic agents. Chief among such agents are treatment measures by anti-convulsant drugs (chemical inhibition), and preventative measures by means of non-convulsive

Table 3.7 Kindling mechanisms and chemicals that affect each

Mechanism	*Chemical effect*
Convulsive mechanism	Effect on rate of development
Retard	Interanimal retardation factor
	Taurine
	Atropine
	Δ^9-tetrahydrocannabinol
	Phenobarbitol
	Diazepam
Facilitate	Reserpine
	6-hydroxydopamine
	Handling (reducing stress and lowering level of norepinephrine?)
Triggering mechanism	Effect on trigger
Suppress	Δ^9-tetrahydrocannabinol
	Phenobarbitol
	Diazepam
	Acetazolamide
	Lidocaine
	Methamphetamine
	Footshock (increase levels of norepinephrine?)
Potentiate	Convulsion producing chemicals (e.g. metrazol)

From Gaito, J. (1976) 'The kindling effect as a model of epilepsy', *Psychological Bulletin*, 83, 1097–109.

electrical stimulation of the brain (electrical inhibition).

Table 3.7 lists groups of chemicals that retard, facilitate, suppress or potentiate convulsions and thus provide models to follow or to avoid in the synthesis of pharmacological agents for the control of clinical epilepsy. Brain stimulation seems a less likely therapy for clinical epilepsy than does the use of drugs. Nevertheless, with the kindling preparation Gaito (1979) reports that under some conditions successive or simultaneous low-frequency (3Hz) stimulation can inhibit an established normal effect (60Hz) either by reversing the elicited behaviour from Stage III to Stage I or II, or by necessitating higher current intensities at the 60Hz frequency for convulsions to be produced.

FOR THE MOLECULAR STUDY OF LEARNING Kindling and learning share common characteristics. They are both relatively permanent effects of experience that must involve the occurrence of changes across synaptic junctions, and they both show positive transfer effects and retroactive interference (Gaito, 1974). As such, Gaito suggests that the kindling effect might provide a paradigm

Behavioural	Normal behaviour			Behavioural automatisms	Clonic convulsions
	Stage 1b			Stage 2b	Stage 3b
Neuro-physiological	Normal activity pattern	Lowering of threshold for AD	1st AD	AD modification	Spread of AD to contralateral amygdala
	Stage 1n	Stage 2n	Stage 3n	Stage 4n	Stage 5n
	Unilateral				Bilateral
Chemical	? ?	?	?	? ?	? ?

Figure 3.1 Illustration of parallel behavioural and neurophysiological events during kindling with unilateral amygdala stimulation. First AD (after discharge) is a simple 1 cps spike and wave form with mean duration of 17 sec and mean amplitude of 702 mV (Stage 3n). In Stage 4n AD changes from a simple to a complex 3–5 cps wave of 10–100 sec duration and mean amplitude of about 986 mV. In Stage 5n ADs appear in contralateral amygdala usually in the form of complex 5 cps waves with a mean duration of 102 sec and mean amplitude of 986 mV. Spikes from contralateral amygdala increase in amplitude until they almost equal those of the ipsilateral site, at which time clonic convulsions occur. (From Gaito, J. (1974) 'The kindling effect', *Physiological Psychology*, 2, 45–50. Reprinted by permission.)

for examining neurophysiological and chemical accompaniments of behavioural changes that occur during learning. Some suggestive changes that have been found during kindling acquisition trials are summarized in Figure 3.1, where the following sequence of electrophysiological accompaniments of the emergence of clonic convulsions with repeated amygdala stimulations is detectable.

1 A normal pattern of electrical activity (EEG).
2 Decrease in after-discharge threshold.
3 An after-discharge that precedes the appearance of stereotypy.
4 Increase in frequency, duration, amplitude and complexity of the after-discharge with the appearance of stereotypies.
5 Appearance of after-discharge in the contralateral amygdala.
6 Convulsions when ipsilateral and contralateral after-discharges are about equal.

Behavioural signs of relatively permanent organismic changes through kindling are four: first, the establishment of convulsions takes several trials; second, kindled convulsions continue for some seconds after the stimulating current is turned off; third, like re-learning or re-conditioning, once-kindled seizures occur with fewer trials on re-kindling than were required for original kindling; and fourth, the appearance of spontaneous seizures is the more likely the greater the amount of original kindling experience. Beyond these, analogously with forgetting, greater current intensity and a larger number of stimulating trials are required for rekindling seizures the longer the period of rest from the experimental situation.

IN ALCOHOL WITHDRAWAL The alcohol withdrawal syndrome, whether in animals or in humans, normally appears only after extensive experience with alcohol. In the study by Falk *et al.* (1972), for instance, rats were exposed to alcohol for over two months before audiogenic seizure tests for withdrawal were made. However, experimental seizures have been obtained after single exposures to alcohol in mice by McQuarrie and Fingl (1958) by use of electro-convulsive shock, and in rats by Mucha and Pinel (1979) with the kindling procedure.

In three experiments with over 30 rats, Mucha and Pinel used 1-second, 60-Hz, 400-μA stimulations of the amygdala and obtained seizures characterized by facial movements, head nodding, forelimb clonus and rearing. The rats were kindled first for three weeks with three stimulations per day spaced half an hour or more apart, and then once a day until reliable seizures were obtained. After that, animals were assigned to groups that received injections of either alcohol (2.0 g per kg rat weight or 2.5 g/kg) or saline solution in conjunction with kindling tests. Mucha and Pinel found that half an hour after the single alcohol injection shorter seizures than normal occurred, but that seizure durations recovered with time after the injection. In all three experiments, longer than normal seizures occurred on kindling tests from 8 to 14 hours after the alcohol injection. This finding is consistent with an early report (Lennox, 1941) on the triggering effect of alcohol on epileptic individuals, and bears on the possibilities that kindling might provide a model for human epilepsy, as well as the possibility that physical dependence on alcohol need not require extensive acquaintance with the drug.

Photogenic seizures in baboons

In human epileptics clinical seizures are sometimes triggered by intermittent light stimulation. The triggering may occur through the accidental play of sunlight on snow or rippling water, by the rhythmic interruption of light from natural or artificial sources, or by close-up viewing of television in the dark (Gastaut and Tassinari, 1966).

A similar phenomenon has been found in baboons, particularly *Papio papio* from southern Senegal. When these animals are exposed to intermittent light stimulation of, say, twenty-five flashes per second while harnessed in laboratory restraining chairs, seizures begin as small eyelid movements that are rapidly followed by diffuse movements in the neck and face, and then tonic–clonic seizures that may be self-sustained after stimulus termination. The self-sustained seizures comprise tonic–clonic phases wherein:

> During the tonic phase, the eyes are open, the pupils dilated, the teeth are bared in a grin, the jaws clenched or half open, the head extended, the arms adducted against the thorax, the forearms flexed on the upper arms, the fists clenched or flexed on the forearms, the thighs in semi-flexion on the abdomen, the legs flexed, the feet in dorsal flexion, and the big toes in planta flexion. (Naquet and Meldrum, 1972, p. 377)

Following this, in the clonic phase, there are:

> Saccadic closing of the eyes, opening or increased opening of the jaws, large jerks of the upper limbs . . . and flexor jerks of the lower limbs. The frequency and amplitude of the clonus diminishes gradually, saliva runs from the corners of the mouth, and then suddenly all motor activity ceases and the

> animal enters the postdictal state of confusion. . . . A few seconds later, the animal makes a few disorganized jumps . . . and . . . then seems to 'wake up'. . . . Recovery usually takes a few minutes. (Naquet and Meldrum, 1972, p. 377)

Reactions of this severity are not shown by all experimental subjects and in a single animal seizures vary in duration from day to day and in occurrence from time to time. From comparisons with other baboon and monkey species, Naquet and Meldrum conclude that this excessive photosensitivity is peculiar to a group of *Papio papio* from the Casamance region of southern Senegal and is not a characteristic of baboons in general. From this it is to be expected that photogenic seizures will have a genetic basis whenever they occur in humans.

4 Behavioural anomalies and misdemeanours

Anomalous behaviour in appetitive situations

Behavioural anomalies must, as Davis (1979) observes, imply behavioural expectations. They are therefore exceptions that prove (challenge, not confirm) the rule and so are suitable bases from which to assault commonsense or theoretical dogma. Freud took the opportunity to mount such an assault on rational psychology through the anomaly of human neurosis; similar opportunities have been taken by students of animal behaviour through the phenomena of displacement activities, animal misbehaviours and normal misdemeanours. By misdemeanours I refer to phenomena that are surprising from the standpoint of current laws of behaviour but that are nonetheless orderly in specifiable environments. The most robust of such misdemeanours in modern psychology are autoshaping, schedule-induced polydipsia and aggression, and barholding with negative reinforcement.

Displacement activities and misbehaviours

DISPLACEMENT ACTIVITIES Displacement activities are apparently irrelevant activities that ethologists have observed in free-ranging animals that are disturbed in the course of feeding, courting, mating, fighting or similar species-specific activities (Tinbergen, 1952). Barnett (1955) has likened them to human psychosomatic disorders. He gives some examples of displacement behaviours in rats:

> A wild rat feeding, disturbed by another rat of a different species, makes 'aggressive' noises and motions towards the interloper; the latter flees; the first rat then, instead of immediately resuming feeding, briefly grooms its face with its forepaws – an action which it would not ordinarily perform in the middle of feeding.

and, during copulation by wild rats:

> At intervals both male and female may make a rapid move to a feeding point, feed hurriedly (and in a quite atypical manner) for a few seconds, and then resume the usual pattern of courtship and coitus. (Barnett, 1955, p. 1203)

Displacement behaviours like these appear mysterious and unlawful in the sense of not fitting to the circumstance at hand, but they force a deeper level of analysis. As Hinde (1962) explains, animals are not single-minded, they do not bring just one behaviour to a situation but many, which must exist in some orderly sequence of readiness. A rat that is eating might otherwise be grooming, but as eating interferes with grooming, grooming cannot begin until eating is interrupted – regardless of the source of interruption. Likewise, copulation inhibits feeding, which cannot occur until sexual behaviour stops. A hungry sex-starved organism can only do one thing at a time, and when one form of action is interrupted the other is likely to occur.

This plausible theory of displacement activity is, however, only one of several (Hinde, 1970). Falk (1972), for instance, considers displacement activity to belong to a class of so-called adjunctive behaviours, which he defines as 'behavior maintained at high probability by stimuli whose reinforcing properties in the situation are derived primarily as a function of schedule parameters governing the availability of another class of reinforcers'. The prototypical adjunctive behaviour from which Falk's definition derives is schedule-induced polydipsia, which is discussed in some detail below.

ANIMAL MISBEHAVIOURS 'The misbehavior of organisms' is an article by Breland and Breland (1961) whose title parodies Skinner's *The Behavior of Organisms* (Skinner, 1938). In it, and in a later book-length elaboration (Breland and Breland, 1966) the Brelands give several illustrations of animal behaviours that do not conform to the empirical law of effect inasmuch as they weaken rather than strengthen with experience. An animal misbehaviour in this respect is the case of the miserly pigs.

> Pigs who have been conditioned [with food reinforcement] to put large metal or wooden dollars in a piggy bank start out doing this neatly and with great precision. They pick up a dollar, put it in the bank, pick up another one, deposit it, and so on. But, after several weeks of performance, the behavior will gradually begin to change. The pig will root the coin, mouth it, toss it, and experience a great deal of difficulty and delay in depositing the coin in the bank. (Breland and Breland, 1966, p. 67)

Such activities are not misbehaviours in actuality, of course, any more than displacement activities are psychosomatic disorders. They are, however, misbehaviours from the standpoint of elementary behaviourism, just as displacement activities are irrational in the eyes of a rational observer. In both cases the lesson is that the behaviour of an organism is not the result of its reinforcement history imposed on a *tabula rasa*, but the result of that history imposed on a selected genetic endowment. This lesson represents an advance in behaviour theory since the original promulgation of behaviourism, and the advance is indebted to some apparently abnormal animal behaviours.

Automaintenance and sign-tracking

A phenomenon that challenges normal learning theory interpretations of behaviour is sign-tracking (Hearst and Jenkins, 1974). Sign-tracking, originally called autoshaping, is a characteristic behaviour of pigeons reinforced on a fixed-time schedule with a signal announcing the impending delivery of food.

Brown and Jenkins (1968) discovered that when the response key in a typical operant conditioning chamber is illuminated shortly before the free presentation of food, naive pigeons, pretrained to eat from the feeder, soon begin to peck the key when the light comes on. The procedure resembles classical conditioning: during feeder training, the food (US) elicits pecking (UR), and then, after autoshaping, the keylight (CS) elicits pecking (CR) which is non-contingently reinforced by food. The unusual feature *vis-à-vis* classical conditioning is that conditioned pecking is directed to the keylight instead of to the food. This feature supports the notion that a Pavlovian conditioned stimulus is a *substitute* for an unconditioned stimulus, rather than that it is a signal that the unconditioned stimulus is imminent, which is the more common view.

An additional peculiarity of autoshaped keypecking is that the behaviour continues even if it prevents food delivery, although there is a topographical difference between pecks that do and pecks that do not cause food to be omitted. Thus, Schwartz and Williams (1972) found, among other things, that under a negative automaintenance regime in which keypecks prevent food-delivery, pecks are shorter in duration than they are under a positive automaintenance regime in which pecks do not affect the food delivery schedule. It is as if the keylight–food association that occurs in the absence of pecking makes future pecking more likely on the basis of classical conditioning mechanisms, and that the keylight–peck response that postpones food delivery inhibits future

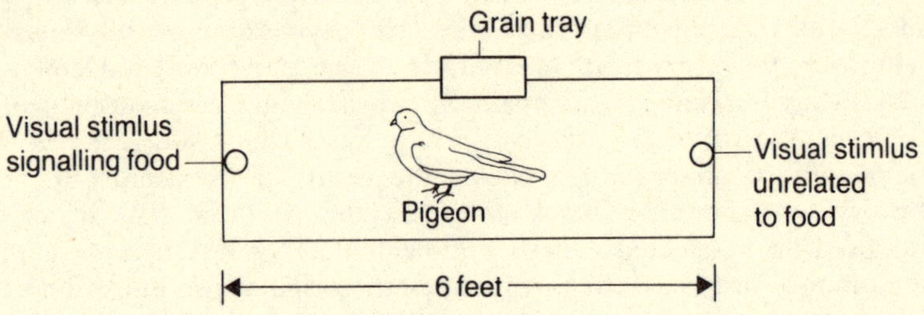

Figure 4.1 Apparatus for studying sign tracking with a remote stimulus. Onset of the visual stimulus at the left predicts brief access to food in the grain tray; the visual stimulus at the right is uncorrelated with food deliveries. The bird acquires the behaviour of approaching and pecking the left stimulus, although it thereby frequently fails to return to the grain tray in time to secure food. (From Jenkins, H.M. (1975) 'Behavior theory today: A return to fundamentals', *Mexican Journal of Behavior Analysis*, 1, 39–54. Reprinted by permission.)

pecking according to an operant conditioning model.

Similar to the negative automaintenance procedure is an experimental arrangement that employs a 'long box' like that shown in Figure 4.1. With this arrangement, Jenkins (1975) found that when a stimulus at one end of the box (the left, say,) reliably predicts a 4-second presentation of food in the tray in the middle, while a stimulus at the right-hand end occurs at random, pigeons soon approach and peck the left-hand signal. As the signal is 3 feet from the food tray, by the time the bird returns to the tray the food has disappeared and the bird goes hungry. 'Nevertheless,' reports Jenkins, 'the birds persist in approaching and pecking the signal.' However the birds cannot persist forever without perishing, and the behaviour, which is abnormal and maladaptive in individual instances, must exhibit a combination of excitatory and inhibitory processes over the long run, as in the case of negatively automaintained responses.

The abnormality of negatively automaintained behaviour is a challenge to the laws of operant and respondent conditioning and helps to refine stimulus, response and reinforcement relationships as they apply to the control of normal behaviour. As Jenkins (1975) concludes:

> The effect of a reinforcer depends on a context. It is not enough to know that a certain movement or stimulus was followed by a reinforcer. It is necessary to learn how to incorporate the setting of that event within an extended interval. Moreover, a reinforcer occurs in an immediate

> context that includes multiple stimuli bearing different relations to the reinforcer. We have to do more than recognize multiple determination, we have to learn the rules. (Jenkins, 1975, p. 40)

And when these rules are known they will alter behavioural expectations such that what now appears anomalous will then be understandable and, if need be, modified.

Schedule-induced polydipsia

Schedule-induced polydipsia is a condition generated in some animals when they secure small pellets of food at short fixed or variable intervals. The phenomenon occurs in several species but is most evident in laboratory rats, with whom it was discovered by Falk in 1961. In Falk's (1961) study, rats consumed on average 92.5 ml of water, which is over three times their normal 24-hour intakes, in experimental sessions that were just over three hours long. The phenomenon is recognized as a potential model of human alcoholism (Gilbert, 1976), as a basis for the provenance of operants (Segal, 1972), and as a challenge to physiological theories of drinking that appeal only to homeostatic mechanisms. In addition, spaced feeding has been employed to induce oral ingestion of addicting drugs (Thompson, Bigelow and Pickens, 1971) and to investigate the potential of anticholinergics and other drugs in controlling overdrinking. Finally, schedule-induced polydipsia is a possible model of psychogenic polydipsia in humans, i.e. polydipsia that is not in response to the polyuria caused by diabetes insipidus.

AS A MODEL OF ADDICTION The role of schedule-induced polydipsia in studies of alcoholism is discussed in Chapter 6. In addition, reference was made to alcohol withdrawal in rats evidenced by audiogenic and kindled seizures in Chapter 3. In both these instances, emphasis is laid upon consumption of alcohol *per se*. Lester and Freed (1972) do not regard this as a suitable criterion for taking schedule-induced alcohol consumption as analogous to human alcoholism, on the grounds that humans drink alcohol for intoxication and anxiety-reduction while rats drink it for its calorific value alone. However, after reviewing the evidence Colotla (1981) concludes that polydipsia is a suitable model of human alcohol overconsumption.

Gilbert (1976) employs schedule-induced polydipsia as a model of drug addiction in an unusual way. In his case, addiction is considered first as excessive behaviour, with the substances of addiction taking a secondary role. This point of view contacts another form of schedule-induced behaviour – aggression – where schedule-induced aggression takes a more violent form than aggression that is directly and explicitly reinforced (Reynolds, Catania and Skinner, 1963).

Spaced feeding generates excessive alcohol consumption when there is no other fluid available, but if rats have the choice of drinking water or alcohol during spaced-feeding experimental sessions, water is the fluid of choice. Keehn and Coulson (1975) found, for instance, that rats who chose 9 per cent (v/v) alcohol (ethanol) over water in their normal home cage drinking reversed this choice for excessive drinking in scheduled-feeding training. Only when the choice of alcohol over water resulted in a higher rate of feeding was more alcohol drunk than water. These authors concluded that schedule-induced drinking alone is not a suitable model of human alcoholism but that it must be supplemented by some other variable that determines the fluid of choice.

FOR DRUG SELF-ADMINISTRATION Animals self-administer drugs in nature incidentally through the ingestion of psychoactive plants (Siegel, 1979). This oral route is supplemented in the laboratory by intubation and inhalation, as well as by intraperitoneal, intravenous and intracerebral injections. In these cases, substances that the animals might otherwise avoid are usually employed. With acceptable substances, oral self-administration permits the study of substance preferences, normally in the form of fluids offered in two- or three-bottle choices. In such choices, some animals exhibit drug preferences, but without excessive consumption or intoxication. With spaced feeding, however, because excessive amounts of fluid are consumed, intoxication is obtainable by means of repeated low doses of drugs that would be unacceptable in single higher concentration doses.

Schedule-induced polydipsia is, for two reasons, especially convenient for the study of alcohol addiction. First, drinking is the normal human route of alcohol intake as a drug of addiction, and, second, overconsumption is the root of the human alcohol problem. So, whether or not schedule-induced overdrinking models human alcoholism, it provides a technique for studying effects of alcohol under self- rather than forced-administration. This is particularly valuable in the study of tolerance, because tolerance is conditionable to the means of administration (Hinson and Siegel, 1982).

IN THE NEUROPHYSIOLOGY OF DRINKING When the spaced feeding procedure induces drug self-administration, the drug naturally has its effects on behaviour, including drinking. Alternatively, drugs have been employed directly for the analysis of schedule-induced polydipsia. In this case the effect of independently administered (usually by injection) drugs on polydipsia is the subject of investigation. Results of some early studies that employed antidiuretic, anticholinergic, stimulant and depressant drugs are summarized in Table 4.1. In an additional study with amphetamine, Colotla and Beaton (1977) report reductions in polydipsic drinking after low-dose (0.5, 1.0, 1.5 and 2.0 mg/kg) injections, and that this occurs because injected animals take fewer and shorter drinks than normal. This result suggests that the

Table 4.1 Pharmacological agents employed in schedule-induced polydipsia research

Drug	*Dosage*	*Effect (IP injections)*	*Investigators*
Hydrochlorothiazide	8 mg/kg	No effect on either water intake or body weight	Falk (1964)
Pitressin tannate in oil	240 mV	Intake reduced, edema fluid rapidly accumulating	Falk (1964)
Phenobarbital	2 mg/kg	Post-pellet licking eliminated; bar-press unaffected; free water intake increased	Falk (1964)
Methamphetamine	0.5 mg/kg	Post-pellet licking eliminated; bar-press unaffected; free water intake decreased	Falk (1964)
dl-amphetamine	0.5 mg – 2.0 mg/kg	Frequency of drinks unaffected	Segal, Oden, and Deadwyler (1965)
Trihexyphenidyl (Artane)	1.5 mg/kg	Attenuated intakes of polydipsic but not non-polydipsic rats	Keehn and Nagai (1969)
Atropine sulphate	0–9 mg/kg	SIP attenuated – greater decrease than with atropine methyl nitrate	Burks and Fisher (1970)
Atropine methyl nitrate	0–9 mg/kg	SIP attenuated	Burks and Fisher (1970)

From Christian, W.P., Schaeffer, R.W., and King, G.D. (1976) *Schedule-Induced Behavior*, Montreal, Eden. Reprinted by permission of the authors.

attenuation of schedule-induced drinking caused by amphetamine may not be a direct effect of the drug on appetite, but an indirect effect of amphetamine-induced fragmented actions (see page 57).

Attenuation or elimination of schedule-induced drinking has been demonstrated with chlorpromazine in the chimpanzee (Byrd, 1974) and chlordiazepoxide in the rat (McKearney, 1973). The anti-psychotic compound haloperidol, administered orally, reduces schedule-induced water consumption by rats in direct relation to dose level, but neither propranolol nor imipramine has any such effect (Keehn, Coulson and Klieb, 1976).

The origination of these drug studies lies in the comparison of schedule-induced with normal drinking. Schedule-induced drinking is unlike normal drinking on account of its excessive nature, but it may not be abnormal in other respects. Although schedule-induced polydipsia is abnormal with respect to the cessation of drinking, it may not be abnormal with respect to its onset or its neurophysiological basis. Normal drinking is initiated by water deficit, but cholinergic drugs like carbachol and acetylcholine elicit drinking by satiated rats when small quantities are placed directly into the hypothalamus or other parts of the Papez circuit of the brain (Fisher and Coury, 1962). This effect is blocked by prior injection of the anticholinergics atropine and scopolamine (Grossman, 1962; Levitt and Fisher, 1966; Stein, 1963). Drugs of this class also attenuate schedule-induced polydipsia (Burks and Fisher, 1970; Keehn and Matsunaga, 1972), suggesting that pharmacologically-induced and behaviourally-induced drinking have similar neurophysiological bases. Likewise schedule-induced polydipsia is reduced by haloperidol, a drug that also controls excessive drinking induced by angiotensin (Fitzsimons, 1973).

IN THE PROVENANCE OF OPERANTS Operant behaviour is, by definition, behaviour that occurs without apparent exteroceptive stimulation. It is controlled by its consequences, not by eliciting stimuli in the normal way of reflexes. In that case, why do operants occur in the first place?

Segal (1972) suggests five possible origins of operants: approximation training (shaping), deprivation, reflex elicitation, releasing stimuli, and emotional induction. In the emotional induction class Segal includes pica (eating inedible substances), air-licking, tail gnawing, wheel-running, and the schedule-induced behaviours of aggression and polydipsia. This suggestion involves two assumptions about schedule-induced drinking: first, that it is initiated by emotion; and, second, that it either is, or becomes, an operant under the control of its consequences. The second assumption is contrary to most data and opinion (see page 85), but Segal (1972) supports it with an account of drinking by a rat food-reinforced on a fixed-interval schedule.

> Midway through the 21st session of the experiment (a fixed-interval 2-minute session) the drinking pattern started changing. A few licks happened to occur shortly before the 43rd food pellet arrived, and then more and more intervals induced licking just before the fixed interval elapsed. . . . By the end of the session, an episode of accelerated-rate licking occurred toward the end of almost every fixed interval. . . . By the 37th session of the experiment, polydipsic drinking had disappeared and been wholly replaced by a fixed-interval-like operant that was apparently controlled by adventitious reinforcement. (Segal, 1972, p. 15)

Less direct evidence that schedule-induced drinking may be operant (maintained by reinforcement) is that drinking facilitates food intake by hungry rats such that food plus water is a superior reinforcer to dry food alone (Keehn and Riusech, 1979), and that drinking diminishes if lick-contingent delays are imposed on opportunities for feeding (Keehn and Stoyanov, 1983).

Segal's first assumption is tenuous insofar as emotion is difficult to define. Nevertheless, to the extent that the Estes–Skinner conditioned suppression procedure operationally defines emotion (see Blackman, 1972; Millenson and de Villiers, 1972, for reviews) then Segal's assumption is upheld by Gillen and Keehn (1983). In their experiment, polydipsic rats maintained on a fixed-interval schedule of food reinforcement for barpressing showed concurrent suppression of barpressing (the index of emotion) and elevation of drinking in periods of exposure to a stimulus previously paired with electric shock.

IN THE ANALYSIS OF BEHAVIOUR The original Falk (1961) discovery of schedule-induced (or psychogenic) polydipsia was two-fold; it involved not only excessive drinking but also food-entrained drinking, that is, overdrinking occurred as a direct result of a hungry animal taking a small drink after nearly every pellet that it earned. The cumulative records of schedule-induced drinking in Figure 4.2 illustrate this effect. Spurs on the records mark occasions of pellet deliveries, and in almost every instance a pellet is followed by a burst of licks. Drinks do not occur in advance of pellets, so there is no opportunity for licking to be maintained by adventitious reinforcement, which was once proposed as an explanation of schedule-induced polydipsia.

Figure 4.2 shows that, regardless of the abnormality in fluid consumption induced by scheduled feeding, schedule-induced

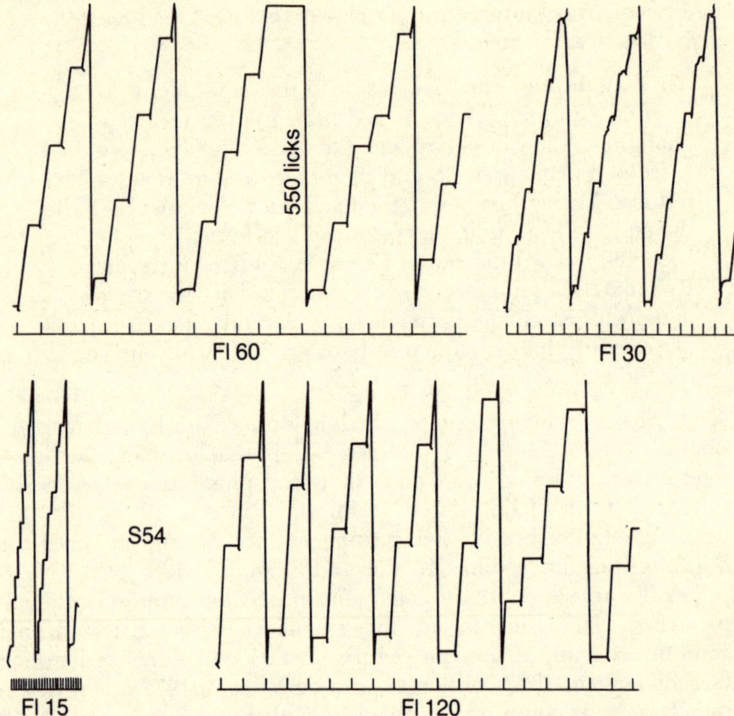

Figure 4.2 The records show post-pellet bursts of licks at a water-tube by a rat (S54) that was rewarded with small (45 mg) pellets of dry food at fixed intervals of 15, 30, 60 and 120 sec as marked on the horizontal time lines. The cumulative recording pen moved horizontally with time and vertically with each lick. Each excursion of the pen marked 550 licks, and each spur on the records marks a pellet delivery. These cumulative licking records are typical of those generated by rats that become polydipsic under intermittent feeding schedules. (From Colotla, V.A., Keehn, J.D and Gardner, L.L. (1970) 'Control of schedule-induced drink durations by interpellet intervals', *Psychonomic Science*, 21, 137–9. Reprinted by permission.)

drinking is a reliable, lawful behaviour. As such it is amenable to analysis in the same way as behaviour specified for reinforcement on a schedule (schedule-specified behaviour). What is mysterious about schedule-induced polydipsia as an abnormality of excess is not a mystery with food-entrained drinking, because drinking after eating is not an unusual phenomenon. From the entrainment standpoint, post-prandial drinking to sluice a dry mouth has been proposed as an explanation of polydipsia. This theory (the dry mouth theory), however, is no more satisfactory than that of adventitious reinforcement in explaining polydipsia, because rats drink after wet as well as after dry food and because they do not drink while food is still available. It thus appears that schedule-induced drinking is not initiated by food in the mouth directly, but by a signal that more food is temporarily not forthcoming.

And drink durations are, up to a point, positively correlated with interpellet intervals. Beyond one- or two-minute intervals, however, post-pellet drinking declines, producing a bitonic relationship between fluid intake and interpellet interval. Figure 4.3 shows how this relationship occurs. At interpellet intervals below about 2 minutes, drinking follows eating quickly and reliably, and drink bursts and intervals between pellets are positively related, but beyond 2-minute interpellet intervals, drinks no longer reliably

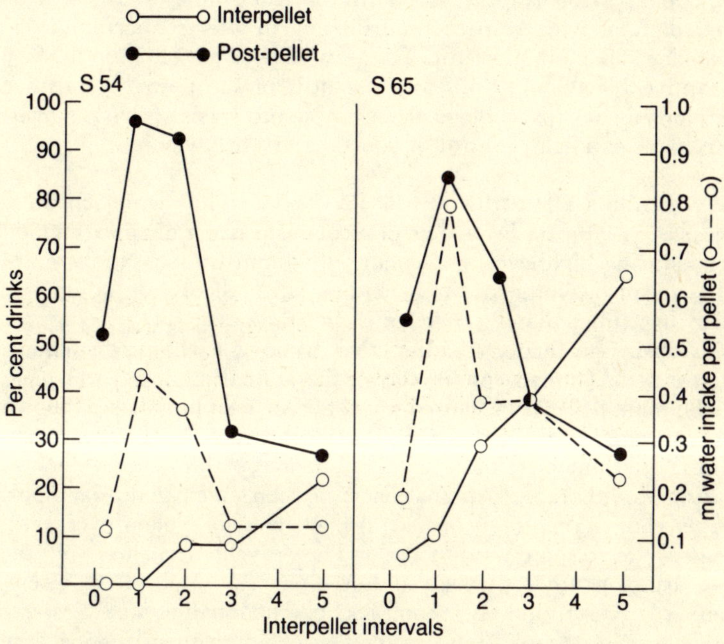

Figure 4.3 Median fluid intakes per pellet, and median percentage of interpellet intervals in which drinking occurred within 10 seconds of eating (post-pellet drinks) and at least 10 seconds after eating (interpellet drinks) over the final three sessions with food pellet deliveries 0.25, 1, 2, 3 and 5 minutes apart (interpellet intervals). (From Keehn, J.D., and Colotla, V.A. (1971) 'Schedule-induced drinking as a function of interpellet interval', *Psychonomic Science*, 23, 69–71. Reprinted by permission.)

follow pellets, and when they occur they occur nearer the ends than the beginnings of interpellet intervals. Thus the entrainment of drinking by eating entails two limiting conditions; the amount of food available without restriction, and the length of the restriction period. If food and water are available freely and concurrently a hungry rat drinks neither regularly nor excessively, but, given the right sizes of meals and the right intervals between them, eating and drinking become entrained and polydipsia ensues.

Polydipsia is a condition of overdrinking, but its main controlling variable is hunger, not thirst. Drink frequencies, durations and quantities depend directly on food (but not on water) deprivation, on the extent to which food is predictable, and on whether or not drinking postpones deliveries of food. As hungry rats eat more food with water available than without it (Keehn and Riusech, 1979), schedule-induced drinking may be biologically functional with respect to eating at the same time as it is maladaptive with respect to drinking. In this connection over-drinking, like obesity, appears as a side-effect of an otherwise adaptive behaviour. Both are instances of the power of positive reinforcement (psychological) over homeostasis (physiological) mechanisms in the control of consumatory behaviours.

AS A MODEL OF PSYCHOGENIC POLYDIPSIA The role of schedule-induced polydipsia in models of alcoholism is mainly as a method for inducing high levels of alcohol self-administration. No theory of alcoholism is implied, and neither is a theory of adjunctive drinking immediately relevant. In the case of psychogenic polydipsia the theory of adjunctive drinking is the direct linkage between the human condition and the animal model.

Human polydipsia caused by diabetes insipidus has a known organic origin and a successful chemotherapy. Psychogenic poly-dipsia in humans has neither an acceptable theory nor a satisfactory therapy. Explanations of psychogenic polydipsia appeal to psychodynamic or behavioural principles to explain overdrink-ing as a response to stress. The psychomechanics of the overdrinking are controversial but all versions take it as stress-induced. With animals, complex psychodynamic analyses are obviously impossible, but an analysis of adjunctive drinking as a stress reaction could possibly be fruitful. From that standpoint, adjunctive drinking would appear in the role of a stress-reducing behaviour, like non-nutritive sucking by infants.

To date there is little evidence of such a function of overdrinking, but promising signs are appearing. In the first place, polydipsic drinking occurs only in hungry animals, and the drinking increases as bodyweight falls. If food deprivation has a general stress-inducing (discomforting) effect as well as the specific effect of inducing hunger, then the adjunctive drinking that occurs with intermittent feeding could be motivated by general discomfort rather than by specific hunger or thirst. And in the second place, there is evidence that in the presence of adjunctive drinking, ACTH blood levels fall (Brett and Levine, 1979).

The stress-reduction theory of polydipsia in animals links psychogenic polydipsia with alcoholism in humans insofar as alcohol is a stress reducing agent, but whereas stress reduction by alcohol can be attributed to its properties as a drug, this obviously cannot be the case with water. Hence it is possible that alcohol consumption and water consumption by non-thirsty organisms both reduce stress via a common orality mechanism, and that the tension-reducing effect of alcohol drinking is as much an effect of the drinking as it is the effect of the drug.

Anomalous behaviour in aversive situations

Schedule-induced aggression

Like schedule-induced polydipsia, schedule-induced aggression occurs as an unprogrammed side-effect of certain reinforcement schedules. The phenomenon is shown by pigeons exposed to schedules with identifiable extinction components (e.g. multiple extinction, variable-interval schedules), and by rats and monkeys subjected to occasional electric shocks.

EXTINCTION-INDUCED AGGRESSION IN PIGEONS Azrin, Hutchinson and Hake (1966) reported extinction-induced aggression in pigeons trained in an operant chamber that contained a companion bird as well as the usual pecking-key and food hopper. Under baseline conditions, with one bird free and the other restrained, no food was presented and no fighting between the birds occurred. Then followed several sessions in which the unrestrained bird was exposed to multiple schedules of reinforcement in which periods of reinforcement and extinction occurred in alternation. With several pairs of birds the free bird forcefully attacked its restricted companion, frequently during extinction periods (and occasionally at other times), even though attacks had no programmed consequences and could lead to food presentations being missed. Similar findings are reported for monkeys and rats.

Like polydipsia, schedule-induced aggression shows a bitonic relationship with interfeeding time, and is excessive in comparison to normal attacks. In fact, so forceful are schedule-induced attacks that experimenters quickly replaced live by mechanical targets in the study of schedule-induced aggression. Flory (1969), for instance, employed a taxidermist's model of a pigeon constructed so that a microswitch closed whenever the head of the model was pecked with a force of 35 grams or more. He presented two pigeons with 4-second access periods to grain at fixed intervals of 15, 30, 60, 120, 240, 480 or 960 seconds, and found that frequencies of attack were bitonic with respect to the interfeeding interval, with a maximum frequency at 60 second intervals for one bird and 120 seconds for the other.

As to its status as an anomaly, extinction-induced aggression is in a comparable position to that of schedule-induced polydipsia; it does not have an obvious function that investigators can agree upon, it can lead to unnecessary physical damage to the aggressor, and it seems excessive in comparison to what might be judged as justifiable aggressive behaviour.

PAIN-INDUCED AGGRESSION Under circumstances in which electric shocks are avoidable by barpresses, rats frequently bite the bar after failing to avoid a shock. However, if two rats are together on occasions of shock, they typically rear into a sparring position and claw or bite each other instead of attacking the bar.

Fighting in respect to pain is possibly adaptive if attack is directed to a persistent aggressor who causes the pain, but it can be maladaptive if an innocent party is attacked. Ulrich (1976) trained rats singly to turn off foot-shock by pressing a bar, and then put them in pairs so that either rat could stop the shock. Two rat heads were not better than one, as it happened, because instead of terminating shocks as soon as they started, the rats fought before one or other of them barpressed and turned off the shock. Logan (1972) reports a similar instance wherein 'a large dominant male rat may hold a small submissive male by the head and prevent him getting to the wheel. . . . And then, when the submissive rat finally turns the shock off, the dominant rat frequently aggresses him again.' The tempting conclusion that rats would rather fight than switch off shock is not supported, however, by a finding of Sbordone, Garcia and Carder (1977) that rats rapidly learn to face away from each other instead of fighting, if facing away serves the purpose of avoiding shock.

With a rubber hose as a target for pain-elicited biting, Hutchinson and Emley (1972) discovered that if a lever is concurrently available, some monkeys press it regularly in inter-shock intervals even though lever presses have no effect. Moreover, the temporal pattern of pressing in the inter-shock interval is the same as that seen when fixed-interval schedules of food deliveries are used to reinforce leverpressing; the rate of pressing the lever is low after shocks, at which times the monkeys vigorously bite the rubber bar, and increases scallop fashion as the next shock time approaches. In this case, leverpressing is not an externally directed attack in the manner of biting and is especially anomalous in that it is maintained by shock just as it would be by food.

Escape and avoidance anomalies

A THEORETICAL ANOMALY Escape and avoidance responses are, respectively, behaviours that terminate and avoid aversive events. They are important behaviours for the study of animal psychology because they are prominent in the contest for survival and because they expose an important difference between respondent and operant conditioning of defence reflexes. The difference is more apparent with leg flexion conditioning studies than with alimentary conditioning.

In Pavlov's case of appetitive conditioning, an original sequence – neutral stimulus, unconditioned stimulus (food), salivation – becomes, after training, conditioned stimulus, salivation, reinforcement (food), such that salivation moves forward from following to preceding the time that food is presented. Pavlov (1927) supposes the same change to occur with the defence reflex.

> The strong carnivorous animal preys on weaker animals and these if they waited to defend themselves until the teeth of the foe were in their flesh would speedily be exterminated. The case takes on a different aspect when the defense reflex is called into play by the sights and sounds of the enemy's approach. Then the prey has a chance to save itself by hiding or by flight. (Pavlov, 1927, p. 14)

The analogy is false, however, because a prey that lives to avoid its predator in future is exposed to Pavlovian extinction, not strengthening. The conditioned flight is not followed by Pavlovian reinforcement – fangs in the flesh – but by avoidance of the predator, which is operant reinforcement.

With Bechterev's conditioned leg flexion method, the procedure is similar to Pavlov's if shock is unavoidable, in which case the consequence of leg flexion is aversive, but if shock is not given after a conditioned response, then the response is operantly, not respondently, conditioned. Because the two types of conditioning were not originally differentiated, avoidance behaviour has been mistaken for conditioned (Pavlovian) escape responding. This mistake also lies in Pavlov's illustration, which could have

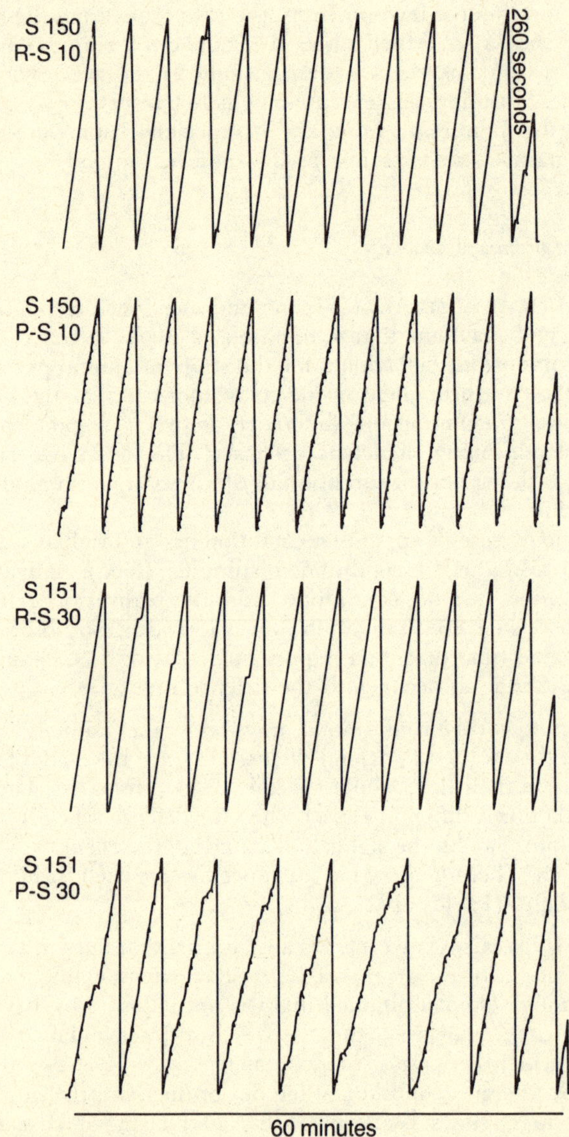

Figure 4.4 Cumulative barholding records of two rats (S150 and S151) trained to barpress to avoid electric shocks. Shocks were delayed either from the time the bar was pressed (P-S) or was released (R-S). Spurs on the records indicate where shocks were not avoided. Whether shock was delayed for 10 seconds or 30 seconds per response, many more shocks were avoided by pressing and holding (R-S) than by pressing and

described a defensive posture instead of a flight reaction, because animal responses to the threat of attack can take such forms as freezing (see below) and fighting (schedule-induced aggression) just as well as fleeing. In fact, Bolles (1971) has proposed species specific defence reactions (SSDRs) as the origins of avoidance learning in laboratory animals, an account that provides an ethological bridge from respondent to operant behaviour.

A popular theory that overcomes the difficulty with the Pavlovian theory of avoidance learning to an extent is Mowrer's (1947) two-factor proposal. According to this, anxiety (fear) is respondently conditioned to a signal of an aversive event, and a response that terminates the signal and avoids the event is operantly reinforced by anxiety-reduction. There are difficulties with Mowrer's theory, similar to those with Pavlov's, but, according to it, avoidance learning by animals is analogous to Freud's second theory of anxiety neurosis. Mowrer's theory is still one of the most popular animal models employed in clinical psychology, and Levis (1979) applies it strenuously in an account of symptom maintenance in anxiety-neurotic humans and of the effects of implosive therapy on the syndrome.

BARHOLDING: A TOPOGRAPHICAL ANOMALY Rats trained to avoid electric shocks by running in an activity wheel or by jumping from a box learn with ease, and the behaviour is well maintained. In double-compartment shuttle-boxes, however, shock avoidance efficiency may decline after learning, and in standard operant barpressing equipment, shock-avoidance may not be learned at all.

The problem with barpresses as a shock-escape or shock-avoidance response is that rats barhold after pressing it terminates shock. Originally, attempts were made to eliminate barholding but later efforts went into analysing its properties (Davis, 1979). Figure 4.4 shows cumulative barholding records of two rats trained with an escape procedure in which shocks occurred at specified times until a barpress occurred, whereupon the shock terminated and the next one was postponed for an interval of 10 or 90 seconds (R–S interval). Usually the R–S interval begins when the bar is pressed, but Figure 4.4 also contains results when the interval starts with bar release. In that case, as the figure shows, shocks hardly ever

releasing (P-S) the bar. The top record shows that S150 received almost no shocks when bar-holding avoided shocks; the one below shows that the animal failed to release the bar, even though many shocks occurred while the animal was bar-holding. The lower pair of records shows the same result with a second animal (S151).

occur because the bar is seldom released, but when barpresses initiate the R–S interval, very few shocks are avoided unless the interval is long, because holding continues until the next shock is received. The behaviour is anomalous because barholding is effortful, and occasional releases and presses would avoid the majority of shocks.

In a detailed topographical analysis of barholding during shock-escape responding, Davis and Burton (1974) uncovered the essential features of the behaviour, as shown in Figure 4.5. In the illustration, a criterion force of 45 grams is required for bar depression, and at A, before shock, the bar is shown depressed. With shock onset, a reflexive lurch leads to increased downward force (B), and then there is momentary release of the bar at C. The animal returns immediately and presses the bar at D, which terminates shock. It continues to hold the bar unsteadily so that spurious barpress responses are recorded at E. After this, barholding above criterion force reappears until the next shock. From such an analysis, Davis (1979) concludes:

> Subjects had learned far less about escape than we had expected; in fact, a virtually untrained animal might perform competently by simply 'doing what came naturally'. On one occasion we placed a deeply anaesthetized rat in the escape chamber, taking care to place him upon the lever in a typical leverholding posture. The subject was incapable of skeletal movement, but the pattern of inter-trial leverholding we recorded on the physiograph was virtually indistinguishable from our 'normal' records. When shock occurred, the rat made a brief reflexive twitch and fell back upon the lever, thereby terminating shock efficiently. (Davis, 1979, p. 203)

This study illustrates not only the lessons for normal psychological theory given by pathological-looking behaviour, but also the danger of automatic transduction of behaviour into computer-printed data without occasional observation, at least, of the living behaving animal.

VICIOUS CIRCLE BEHAVIOUR Vicious circle behaviour in rats derives its name from Karen Horney's (1937) depiction of the vicious circle of self-defeating neurotic behaviour in humans. It is also called self-punitive behaviour, and seen as analogous to masochism. It was first studied experimentally by Mowrer (1947), who noted that rats trained to avoid electric shocks persisted longer in the avoidance response during extinction if shocks were occasionally encountered than if they were eliminated entirely. In a typical reference experiment (e.g. Brown, Martin and Morrow,

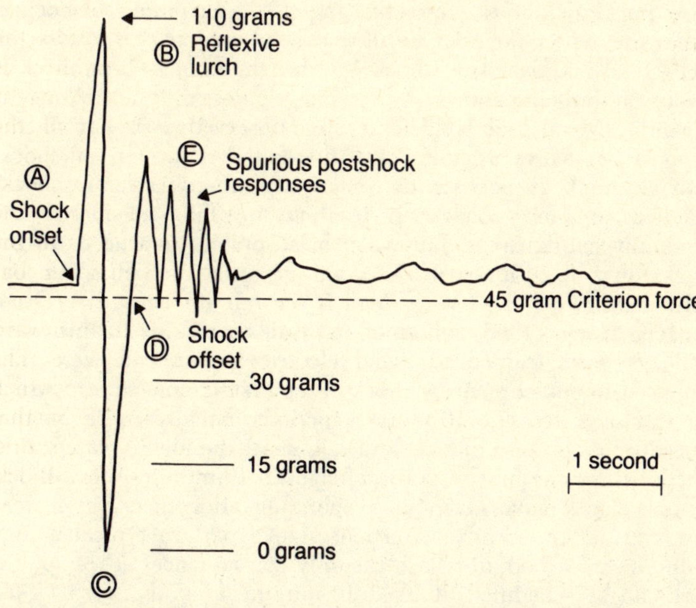

Figure 4.5 Composite illustration of major features of response-force records obtained during initial portions of inter-shock intervals. The rat was exerting about 48g downward pressure at the moment of shock (A). As it lurched from the lever (B) a peak force was momentarily reached, followed immediately by a return to 0g (C) when the rat was away from the lever. When the rat returned to the lever, shock offset occurred when the downward pressure reached the criterion value of 45g (D). Spurious responses are recorded between D and E as pressure on the lever fluctuated above and below the criterion value, after which the rat remained holding the lever down with an above-criterion force until the next shock onset, even though lever-holding was non-functional. (From Davis, H., and Burton, J. (1974) 'The measurement of response force during a leverpress shock escape procedure in rats', *Journal of the Experimental Analysis of Behavior*, 22, 433–40, Copyright 1974 by the Society for the Experimental Analysis of Behavior, Inc. Reprinted by permission.)

1964), rats are trained to escape from shock by running from a start box down an electrified alley to safety in a goal box. After training, shocks are discontinued altogether for some animals, but for others, although shocks are never given in the start box, they are given in the alley if the animals enter it from the start box. The latter animals could remain unshocked in the start box, but instead they continue to enter the electrified alley and to run through it, although more quickly than the unshocked animals. That they run more quickly is due to the motivating effect of painful footshock; it is that they enter the alley at all that is anomalous, for alley-entry is effectively punished by shock – hence the name self-punitive or vicious circle behaviour.

Morse, Mead and Kelleher (1967) observed a similar phenomenon in monkeys. In their case they found that after tail shocks squirrel monkeys persistently pull and bite a restraining leash attached to their collars. If leash-pulling has no function it eventually extinguishes, but when leash-pulling produced electric shocks on a fixed-interval 30-second schedule the behaviour was emitted exactly as it is when food is the reinforcing agent.

McKearney (1968) obtained a similar result. In his case, monkeys were trained to avoid electric shocks and were then subjected to punishment by shock of the first avoidance response at the end of successive 10-minutes periods. Finally the avoidance procedure was terminated and all that remained was shocks scheduled for the first response after each 10-minute interval. The animals could have given up responding altogether, but instead they emitted thousands of responses in a pattern typical of that maintained by food, although the only consequences of responding were shocks scheduled at fixed 10-minute intervals. Byrd (1969) found a similar effect with response-independent shocks in a study employing cats, as did Branch and Dworkin (1981) with response-produced shock by monkeys.

These are anomalous behaviours because they appear to be self-inflicted punishment. Psychologists have an unusual difficulty over punishment, being uncertain about its status as an effect (behavioural suppression) or as a procedure (e.g. presentation of electric shock). Electric shock is, however, just a *stimulus* that can be defined by its physical attributes independently of its effects, and can have different effects at different times, as in the case of food. Food can be specified by quality and by quantity regardless of a consumer, and can either reinforce or nauseate the consumer, according to whether he is hungry or well-fed, or whether the food is familiar, novel or associated with pleasant or painful events. Likewise the aversiveness and behavioural effect of electric shock must be distinguished according to context. In fact, one of the earliest illustrations of 'experimental neurosis' occurred in an experiment where electric shock was employed as a conditioned stimulus for salivation (Yerofeeva, 1916). Possibly the special difficulty we have with punishment is that most people are not as familiar with its opposite effect from normal (pleasure) as they are with the opposite effect from normal with food (nausea). Thus, while pleasure from pain has long been a well-known clinical syndrome, nausea from eating, in the forms of bulimia and anorexia nervosa, has only recently become a prominent psychological problem.

DISCRIMINATIVE PROPERTIES OF PUNISHMENT Salivation to electric shock, as in the above-mentioned Yerofeeva experiment, is anomalous only if the conditioning history of the organism is unknown. With knowledge of that history, both suppressive and facilitative effects of a demonstrable aversive stimulus can be understood, as the following study by Holz and Azrin (1961) shows. The anomaly from the standpoint of a naive observer is that the subject emits a response more rapidly on occasions when the response is shocked than on occasions when shock is omitted. The experiment was conducted in three stages, the first of which served to establish the behaviour that was subsequently punished. This stage is essential in the study of punishment, because any response that requires suppression by punishment must be independently maintained by some source of reinforcement, otherwise punishment would not be required. Under natural conditions the establishing and maintaining reinforcer of undesirable behaviour may be unknown, but in the laboratory this reinforcer can be specified and controlled, as it was by Holz and Azrin.

The experimental subjects were two pigeons. After preliminaries, they were conditioned to peck a key with food reinforcements delivered on a 2-minutes variable interval schedule. With this schedule the pigeons key-pecked at a steady rate of about fifty pecks a minute, receiving some fifteen reinforcements in half an hour. When this behaviour was stable, each peck was punished with an electric shock. This served to halve the rate of responding without materially affecting the number of food reinforcements the animals received. The third step was the introduction of extinction sessions in which both food reinforcement and punishment were withheld. In these sessions response rates declined almost to zero. Thus, there were:

1 sessions in which key-pecks were never punished and sometimes reinforced;
2 sessions in which key-pecks were punished and occasionally reinforced; and
3 sessions in which key-pecks were neither punished nor reinforced.

Only in the last case did response rate fall to zero. Punishment did serve to reduce response output but it was less effective in suppressing responding than was withdrawal of reward. However the most significant finding is shown in Figure 4.6. In this instance no food was given, and punishment was introduced for a short test

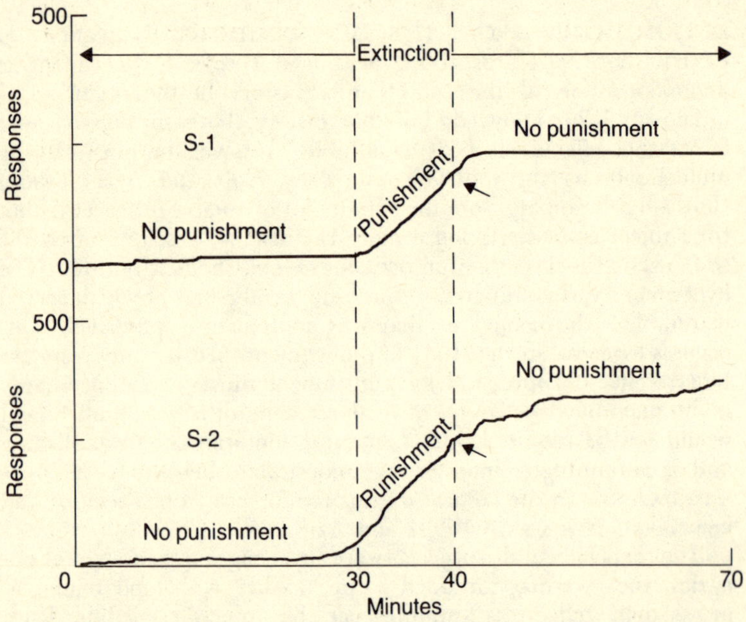

Figure 4.6 Cumulative pecking responses of two pigeons when a 10-minute punishment period occurred between two 30-minute periods without punishment. Punishment had previously been made a discriminative stimulus for food reward, by the correlation of punishment with feeding and non-punishment with extinction. The arrows point to increased response rates that occurred immediately after punishment terminated. (From Holz, W.C., and Azrin, N.H. (1961) 'Discriminative properties of punishment', *Journal of the Experimental Analysis of Behavior*, 4, 225–32, Copyright 1961 by the Society for the Experimental Analysis of Behavior, Inc. Reprinted by permission.)

period in the middle of the session. During this period response rate rose. Instead of suppressing the pecking response, punishment facilitated it. This is an anomalous result according to normal expectations of punishment, but in this case the reason is clear; after their early training the animals were reinforced in the presence of punishment and not reinforced in its absence, a circumstance that establishes a stimulus as a discriminative stimulus for a response. Hence in the absence of punishment response rate was low because no punishment meant no food, and in the presence of punishment it became relatively high because the presence of punishment signalled the possibility of food. As in the Yerofeeva respondent conditioning experiment, the normally aversive stimulus of electric shock is transformed into a signal for food, and so occasions anomalous-looking behaviour to an observer unfamiliar with the training history of the animals.

LEARNED HELPLESSNESS In 1949, H.F. Harlow demonstrated the ability of monkeys to learn how to learn; practice on one problem (intra-problem learning) led to faster learning of similar problems (inter-problem learning). He called this the acquisition of *learning set*. A recent similar discovery is the phenomenon of learned helplessness, where experience with an insoluble problem occasions learned failures in other situations where learning would normally occur. In animals, Seligman (1975) regards learned helplessness as an experimental model of human depression.

An experimental demonstration of learned helplessness is that of Seligman and Maier (1967). They pre-trained three groups of dogs before submitting them to an escape learning procedure in a standard shuttlebox apparatus. In the pre-training, dogs in one group (escape) could terminate electric shocks to the hind legs by pushing a contact plate with their heads; dogs in a second group (yoked controls) received shocks exactly like the dogs in the escape group, but could not control them; and dogs in the third (naive controls) group were harnessed like the others but not shocked. The results of the second part of the experiment, in which all the dogs were trained to escape by shuttling from one to the other compartment of the shuttlebox were that the dogs in the escape and naive groups learned to escape shock in the shuttlebox by jumping more and more rapidly trial after trial, but that no learning occurred with the yoked group that was previously exposed to uncontrollable shock.

Seligman (1975) explains such a result as the learning of a negative cognitive set. A similar result, with rats, is explained by Weiss, Glazer and Pohorecky (1976) as due to neurochemical changes induced by stress. In their experiment, Weiss *et al.* used five groups of rats, all of which were given escape-avoidance training in a standard shuttlebox in which a 5-second warning signal preceded shock. The groups were differentially treated 30 minutes before this training. Two groups were forced to swim in cold water (2 °C) for 6.5 seconds and 3.5 seconds respectively, an experience that depletes brain norepinephrine but presumably does not induce defeatism concerning shuttlebox training; two groups were forced to swim in warm water (28 °C), which neither depletes brain norepinephrine nor induces helplessness; and a control group was given no pretest swimming experience.

The results in terms of response latencies to the warning signal showed that the cold swim prior experience, which differentially depleted brain norepinephrine, markedly reduced escape-avoidance performance, whereas the warm swim prior experience did not. In fact, the 6.5-second cold-swim animals seemed almost incapable of moving in the shuttlebox. The 3.5-second cold-swim group was not so totally incapacitated; they reared up and squealed at shock onset but failed to run from one to the other side of the shuttlebox until between 10 seconds and 25 seconds of shock had occurred. The other groups rapidly crossed compartments after shock onset, and sometimes before the shock came on.

Porsolt, Pichon and Jalfre (1977) similarly found that immersion in water produced temporary immobility in rats, and also that the duration of immobility is reduced by electroconvulsive shock and antidepressant drug treatments. On this account, coupled with the thought that the 'behavioural immobility indicates a state of despair in which the rat has learned that escape is impossible,' these workers independently arrived at an animal model of depression that combines Seligman's 'learned helplessness' model and the neurotransmitter depletion model of Weiss and his colleagues.

Anomalies in practice and theory

Tuber, Hothersall and Voith (1974) have demonstrated some values of a practical clinical psychology of animals, and Hothersall and Tuber (1979) have produced additional illustrations. These illustrations focus on household pets, but the need for a clinical psychology of animals is also evident in problems met by exhibitors of pedigree show animals and by farmers with dairy herds (Seabrook, 1982). In this latter case, it is not so much a question of deviant behaviour by individual animals as of loss of productivity on occasions. Seabrook (1982) has given the question attention and found that the milk yield of a herd is affected by the temperament of the herdsman. The contented cow is more than a convenient slogan, for, by Seabrook's research, the cowman who is a confident introvert elicits more milk from a herd than one who is a talkative, sociable worrier. Given the need for greater food production in underdeveloped nations, the use of agricultural psychology to facilitate improved animal–human interactions could eventually be of enormous practical value.

However the behavioural anomalies described earlier in this chapter have another value – that of putting behavioural theory to the test. Anomalies may draw attention to otherwise easily overlooked behaviour, and so extend the boundaries of normal theory, or they may provide the exceptions that prove normal theory wrong. Eysenck's (1957) extension of normal learning theory to account for abnormal neurosis illustrates the first of these alternatives; Freud's reconstruction of normal psychological theory on account of irrational neurosis is an example of the other (Freud, 1936). The more an established theory can successfully encompass novel eventualities under its umbrella the more assurance do we have that the theory is correct, but inasmuch as science rests on the principle that the only certainty is error, it is through anomalous, unexpected, behaviours that new knowledge is able to emerge.

Part III
Animal models of disease entities

5 Experimental neurosis and psychophysiological disorders

Manifestations of neurosis: human and animal

Neuroses are disabling and distressing human conditions that either have no organic origins, or, if they have, persist long after these origins disappear. The American Psychiatric Association's third *Diagnostic and Statistical Manual* (*DSM-III*) replaced the blanket diagnostic category of neurosis with anxiety reaction and a number of separate syndromes, but different names and different classifications have been adopted from time to time. The two main groups of neurotic reactions have been called neurasthenia and hysteria, although neurasthenia is also called anxiety neurosis, asthenic reaction or dysthymia. For a time, hysteria and neurasthenia were considered to be physical disorders but now they are known to be psychogenic and are often called psychoneuroses to emphasize their psychological rather than their neurological basis.

Neurasthenia refers to nervous exhaustion. A symptom is chronic fatigue and the syndrome includes insomnia, concentration difficulties and anorexia as well as more direct physical aspects like headaches, dizziness and vague aches and pains. Classical cases of hysteria also show physical malfunction, but without a corresponding organic defect. Hysterical disabilities may originate as physical symptoms or be exaggerations of actual physical illnesses. Hysterical reactions may appear as motor disorders, sensory disorders or disturbances in the autonomic nervous system. Asthenic reactions include anxiety states, obsessional and phobic neuroses.

Neurotic personalities may exhibit complex mosaic patterns of these symptoms. A remarkable case of an apparently neurotic dog is the mongrel that Gantt (1944) called Nick. This dog exhibited at one time or another almost every kind of neurotic symptom possible, and he may be compared to Breuer's classical case of Anna O.

> Nick showed a train of symptoms soon after work was begun with him in 1932 and continuing with some intermissions until he was killed in 1941. These consisted of asthenic-like breathing, extreme pollakiurea, ejaculatio praecox, gastric hyperacidity; also visceral perturbations, great restlessness, agitation, negativism towards people he knew who brought out [these] symptoms, plus sexual erections and often ejaculation. Furthermore, he was, paradoxically, impotent in the presence of an adequate stimulus. (Gantt and Dykman, 1957, p. 17)

By Jones' (1961) account of Anna O, whom Breuer treated from December 1880 to June 1882:

> The patient was an unusually intelligent girl of twenty-one, who developed a museum of symptoms in connection with her father's fatal illness. Among them were paralysis of three limbs with contractures and anesthesias, severe and complicated disturbances of sight and speech, inability to take food, and a distressing nervous cough which was the occasion of Breuer being called in. More interesting, however, was the presence of two distinct states of consciousness: one a fairly normal one, the other that of a naughty and troublesome child. It was a case of double personality. (Jones, 1961, p. 147)

Breuer's case is one of a natural neurosis. It helped Freud start the psychoanalytical movement towards a psychological explanation of certain mysterious physical illnesses that stresses as their causes prior mental trauma rather than neurological lesions. Gantt's case is one of experimental neurosis. It followed Russian demonstrations of emotional breakdowns in dogs caused by stresses deliberately induced in the laboratory. The two cases together illustrate by their subsequent histories the complementary contributions of humans and lower animals to the study of neurotic disorder remarked on by Liddell (1956); symptomatology in the case of humans and originating situations in the case of animals.

The multiple personality syndrome of Anna O belongs in a subcategory of neurosis – dissociative reactions – that includes amnesia and fugue states, which are less dramatic and exaggerated examples of memory deficits than are multiple personalities. Anecdotal accounts of possible analogues of such neuroses in farm animals are given by Croft (1951), and cases of 'multiple personality' in dogs are recorded by Scott and Fuller (1965) and Fox (1965). The Fox case is an adult cocker spaniel that was peaceful all day while alone with his mistress but ferocious at night when her husband came home – a classical candidate for an Oedipal interpretation! Scott and Fuller describe a basenji, Gyp: 'Perfectly behaved in the laboratory, he became almost a canine delinquent at home and eventually became uncontrollable.'

These cases cannot seriously be interpreted in terms of symptomatology – Oedipal or otherwise – but their potential for analysis on the basis of origination is obvious. The same can be said of hysterical reactions in dogs (Parry, 1948) and in poultry (Sanger and Hamdy, 1962; Hansen, 1976; Katz, 1937). Sanger and Hamdy report on periodic episodes of mass hysteria in white Leghorn hens that 'was characterized by wild running, piling, crowding under feeders and waterers, hiding in nests, or flying as high as possible to perches above the floor' of the poultry shed. The origin of this behaviour appears to lie in overcrowding (Hansen, 1976), but this is not so of the example given by Katz, who recounts an experiment by Bruckner in which chicks were raised in isolation.

In Bruckner's experiment, freshly incubated chicks were removed directly from the incubator to a warm illuminated box where they were watered and fed in total isolation to study the effects of maturation without training on behaviour. However, according to Bruckner's account, after a week in isolation the chicks began chasing stray flies 'as if possessed'. One chick ran back and forth knocking two or three times at the front and back walls of the box at each visit, and another, when offered a worm, 'seized it with its beak, turned around as though other chicks wanted to snatch the worm away and ran in a circle much the same as if in a real chase' (Katz, 1937, pp.217–18). And yet the chick had always lived alone.

Katz reports these observations under the heading of 'Neurotic behaviour brought about by social isolation'. The behaviour of the isolation-reared chicks resembles repetitive stereotyped activities of dogs and primates reared in isolation, and the resemblance could reveal cross-species behavioural analogues. Apart from the question of whether the abnormal behaviours of dogs, chicks, monkeys and chimpanzees raised in isolation are analogous to each other, there is also the question of whether each or any of them is analogous to a human neurotic condition. The answer rests on what neurosis is supposed to mean.

Meaning of neurosis

Disease entity or clinical pictures?

Two approaches to the meaning of neurosis are by way of theoretical definition and by way of enumeration of symptoms. Hunt (1964) gives an imaginary example of the first approach.

> By definition, a neurosis involves intra-psychic conflict in which ego and superego play a crucial role and in which the dramatic psychic events associated with the Oedipus complex and its resolution lay an essential groundwork.

From this theoretical approach the inevitable conclusion follows: 'Animals have no egos, superegos, or Oedipus complexes – consequently no neuroses, however peculiarly they act.' However, theoretical definitions of neurosis, even psychodynamic ones, need not exclude animal possibilities. Experimental neurosis in animals has been offered as evidence for defining neurosis in each of the following theoretical contexts:

1 as a pathophysiology, or failure of process (Pavlov, 1970);
2 as a form of faulty learning, or failure of the outcome of a normal behaviour process (Wolpe, 1952);
3 as a psychopathology, or failure of dynamic psychic processes (Masserman, 1943).

The second approach to the definition of neurosis simply lists symptoms of classical neurotic syndromes. Thus, Woodruff, Goodwin and Guze (1974) define anxiety, obsessional, phobic and hysterical neuroses as follows.

> *Anxiety neurosis* is a chronic illness characterized by recurrent, acute anxiety attacks having a definite onset and spontaneous termination. During attacks the patient is fearful and has symptoms associated with the autonomic nervous system: palpitations, tachycardia, rapid or shallow breathing, dizziness and tremor.
> *Obsessional neurosis* is an illness dominated by obsessions and compulsions occurring in the absence of another psychiatric disorder. Obsessions are persistent distressing thoughts or impulses experienced as unwanted and senseless but irresistible. Compulsions are acts resulting from obsessions.
> *Phobic neurosis* is a chronic disorder dominated by one or more phobias. A phobia is an intense, recurrent unreasonable fear.
> *Hysteria* is a polysymptomatic disorder . . . characterized by recurrent, multiple somatic complaints [including

> unexplained] varied pains, anxiety symptoms, gastrointestinal disturbances, urinary symptoms, menstrual difficulties, sexual and marital maladjustment, nervousness, mood disturbances and conversion symptoms . . . Conversion symptoms are unexplained symptoms suggesting neurological disease, such as amnesia, unconsciousness, paralysis, 'spells', aphonia, urinary retention, difficulty in walking, anaesthesia, and blindness.

There is no characteristic common to all these definitions of sub-classes of neurosis, which is why only sub-classes are retained in *DSM-III.* The above four syndromes form two distinctly different groups: *anxiety neurosis and hysteria*, which are characterized by autonomic and emotional disturbance; and *obsessional and phobic neurosis*, which are unreasonable and irresistible urges. Neurotic behaviour as a whole is behaviour that is unreasonable, but in the anxiety and hysteria versions reason is overwhelmed by emotion, whereas in the obsessional and phobic varieties it is unreason that replaces reason. Neurosis cannot be proclaimed a clinical entity by this account; it does not have a positive defining characteristic, only a negative one of absence of reason. Thus neurosis is not a disease entity like bronchitis or tuberculosis, but a blanket categorization of imperfection like 'broken', 'unhealthy' or 'disabled'. The syndromes that Woodruff, Goodwin and Guze (1974) describe are not disease entities but clinical pictures, and as such can be exemplified by animals without implication of human psychic processes. Several illustrations of such clinical pictures appear below.

Criteria for animal neurosis

Two other approaches to the definition of neurosis have been adopted, one by Eysenck (1957) with reference to humans and one by Hebb (1947) with reference to animals. Paradoxically, Eysenck's theory of neurotic syndromes stems from Pavlov's laboratory studies with dogs, while Hebb's criteria of neuroses in animals is based on clinical accounts of humans.

Eysenck analyses the dynamics of anxiety and hysteria into two personality dimensions, neuroticism and introversion–extraversion, from which two kinds of neurotic disorders are identified: disorders of the first kind, or introverted neurosis; and disorders of the second kind, or extraverted neuroses. At the symptom level, the first kind of disorder includes anxiety, phobias and disturbed emotions, while the second kind includes psychopathy, homosexuality and fetishism. Originations of the two kinds of neurosis are attributed to animal-laboratory based phenomena; traumatic conditioning for the first kind, and faulty conditioning for the second. In this system, Broadhurst (1973) wonders if a factor of autonomic imbalance identified in animals corresponds to Eysenck's neuroticism factor, and Gray (1964) links the introversion–extraversion dimension with the temperamental types of animals postulated by Pavlov (see below).

By contrast, Hebb (1947) focuses on neurosis in general, and defines it objectively as a generalized and persistent undesirable emotional condition.

> Neurosis is in practice an undesirable emotional condition which is generalized and persistent; it occurs in a minority of the population and has no origin in a gross neural lesion. (Hebb, 1947)

From this definition arise six criteria for attributing neurosis to animals, but none for recognizing neurotic clinical pictures. The criteria of neurosis are:

1 evaluationally undesirable;
2 emotional;
3 generalized;
4 persistent or chronic;
5 statistically relatively infrequent;
6 not due to a specific gross lesion.

These criteria Hebb derived from humans. He recounts two cases of spontaneous neurosis in chimpanzees that appear to meet them. One of the cases is Alpha, a 15-year-old female born and raised in captivity.

> At the night feeding of November 6, 1942, with no earlier sign of abnormality, Alpha refused all solid foods. There was no other sign of illness, and she was clearly hungry next day, in spite of the pieces of food lying untouched in her cage. No coaxing could persuade her to touch the food . . . let alone touch it [she] would make a wide circle around [a large piece of food] that she had to pass in moving about the cage. Four months after the onset of the illness, when the avoidance of food . . . was much less marked, Alpha showed a sudden and marked avoidance of the attendant who had been feeding her. . . . This lasted seven days. On the eighth, there was again fear of food and none of [the attendant]. (Hebb, 1947, p. 4)

The origin of Alpha's specific symptomatology, which includes further alternating bouts of fear of food and the attendant, must remain a mystery. Hebb argues that it does not derive from specifiable noxious events, although recent work on one-trial learning of taste aversions weakens this case (Rozin and Kalat, 1971), but from examples of captive chimpanzees' disorders recounted earlier it is possible to agree with Hebb and appeal to chronic effects of caging as a basis of Alpha's abnormality.

Experimental neurosis

Pavlovian prototypes

DISCRIMINATION AND GENERALIZATION FAILURES Occurrences of human neuroses are scattered through history. The discovery of natural and experimental 'neuroses' in animals may be attributed to Pavlov. In 1924, water from the overflowing Neva seeped into Pavlov's laboratory in Leningrad so that his experimental subjects had to be taken from their quarters to safety. Afterwards it was noticed that dogs that had already formed conditioned responses lost them and became nervous and excitable when they returned to the laboratory. Pavlov equated the dogs' emotional breakdowns with neurosis in humans. Similar breakdowns were created in the laboratory, accidentally by Yerofeeva in 1912 (Erofeew, 1916), and more methodically by Shenger-Krestovnikova a decade later (Pavlov, 1927).

Yerofeeva used electric shock as a conditioned stimulus for the salivary reflex, a not uncommon practice in Pavlov's laboratory, but when she attempted to generalize the response to shocks applied to other parts of the body, instead of salivating according to expectations her animal defensively struggled against the shock. The Shenger-Krestovnikova experiment is the classical one in which salivation was conditioned to a circle but not to an almost circular ellipse. In the face of the difficult discrimination, emotional breakdown occurred after several days of training, as Pavlov's original account of experimental neurosis makes clear.

> The differentiation [between secretions to the circle and not to the ellipse] proceeded with some fluctuations, progressing at first more and more quickly, and then again slower, until an ellipse with ratio of semi-axes 9:8 was reached. In this case, although a considerable degree of discrimination did develop, it was far from being complete. *After three weeks of work upon this differentiation* not only did the discrimination fail to improve, but it became considerably worse, and finally disappeared altogether. *At the same time the whole behavior of the animal underwent an abrupt change.* The hitherto quiet dog began to squeal in its stand, kept wriggling about, tore off with its

> teeth the apparatus of mechanical stimulation of the skin, and bit through the tubes connecting the animal's room with the observer, a behavior which never happened before. On being taken into the experimental room the dog now barked violently, which was also contrary to its usual custom; in short it presented all the symptoms of a condition of acute neurosis. (Pavlov, 1927, p. 291, emphasis added)

PATHOPHYSIOLOGY OF EXCITATION AND INHIBITION From these and similar data Pavlov (1970) constructed a theory of experimental neurosis that appealed to clashes between excitatory and inhibitory neural processes in the brain. These processes are presumed to underlie the basic conditioning phenomena of acquisition, extinction, generalization and discrimination, which are governed by laws of *intensity* (where response magnitude is proportional to stimulus intensity), *irradiation and concentration* (where weak stimuli irradiate throughout the brain from the point of reception, medium stimuli irradiate from and concentrate back to that point, and strong stimuli oscillate), and *reciprocal induction* (where inhibition is introduced at a point of concentrated excitation, and *vice versa*).

In this context, experimental neuroses are taken to be pathological failures of higher nervous processes, exemplified by emotional disorders caused by overstrain of excitation or inhibition, or of the balance or mobility (see below) between them, and by discrimination failures of the following patterns in previously well-trained subjects.

1 *The equalization pattern*, in which conditioned response magnitude becomes the same to weak as to strong conditioned stimuli. This pattern exemplifies a breakdown of the law of intensity, whereby normal conditioned response magnitude is a direct function of conditioned stimulus intensity.
2 *The paradoxical pattern*, in which responses to weak conditioned stimuli are stronger than those to strong conditioned stimuli.
3 *The ultraparadoxical pattern*, in which more conditioned salivation occurs to the unreinforced (differential) conditioned stimulus than to the reinforced conditioned stimulus.
4 *The inhibitory pattern*, in which no salivation occurs to either the positive or the negative conditioned stimuli.
5 *The excitatory pattern*, in which salivation that once occurred only to the positive, reinforced, conditioned stimulus now occurs to the negative, differential, stimulus as well.

These patterns refer only to failures of differentiation of the conditioned response. In addition, as the above quotation from Pavlov shows, much additional behaviour also occurs in experimental neurosis, including disturbances of physiological functions. These latter disturbances are not 'symptom-specific' in the sense that particular stressors produce particular pathological patterns in all dogs, but appear to relate to the temperament as well as to the conditioning history of a particular animal. The temperamental type most susceptible to experimental neurosis is the one that Pavlov associated with the weak nervous system in the following typological system.

ANIMAL TYPOLOGY Pavlov believed that individual susceptibilities to neurotic breakdowns under conditioning procedures were determined by constitutional differences in the strengths of excitatory and inhibitory tendencies, their mobility and their balance in individual cases. Figure 5.1 summarizes his typological system and relates it to the Galenic humoral theory of temperaments (Macmillan, 1963). Dogs with weak nervous systems (melancholics, assessed by complex independent conditioning tests) are said to succumb most easily to experimental neurosis, followed by the strong unbalanced (excitatory, choleric) type, with lively, sanguine dogs most resistant of all to experimental neurosis. When a dog did become neurotic, Pavlov used sedation by bromides to treat the neurosis, and found that strong excitatory dogs required much larger bromide doses to sedate them than did dogs of the weak inhibitory type. Eysenck (1957) and Gray (1964) have applied these Pavlovian discoveries to a theory of human

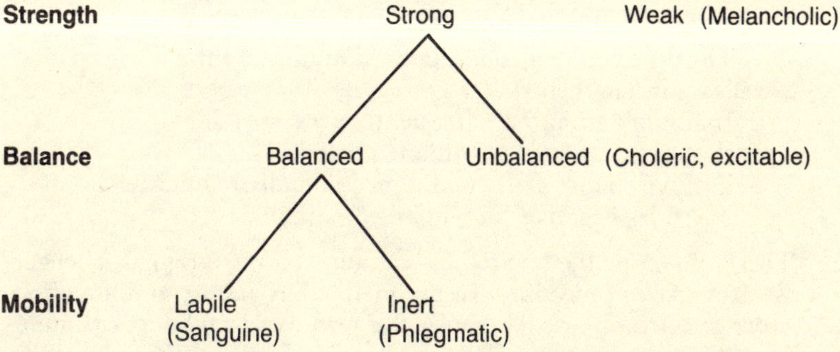

Figure 5.1 Pavlov's typology of nervous systems of dogs, based on the strength, balance and mobility of excitatory and inhibitory higher nervous processes. Bracketed terms are Hippocratic equivalents.

personality, and Reese (1978) has identified two types of purebred dogs that are, respectively, vulnerable and resistant to experimental neurosis. The first, type-E dogs, seem not to behave normally in either operant or respondent experimental situations, and are 'markedly disturbed' even before exposure to experimental conflict situations. The resistant type-A dogs, on the other hand, are reported as normally active, inquisitive and gentle dogs that comply with experimental demands and resist experimental neurosis in both painful and difficult discrimination situations.

Methodological elaborations

BASED ON PATHOPHYSIOLOGY Elaborations on Pavlov's original methodology for inducing experimental neurosis in dogs have involved innovations in experimental techniques and extensions of the variety of species under investigation (Lagutina and Sysoeva, 1969/70). However the underlying theoretical guidelines of Russian experimentation have not deviated from Pavlov's insistence on neurosis as a pathophysiological condition.

All but the most elementary of classical conditioning methods involve, according to theory, excitation and inhibition. The simple pairing of conditioned and unconditioned (reinforcing) stimuli is supposed to invest excitatory properties in the conditioned stimulus by which the conditioned response is elicited. However, during time intervals between onsets of the paired conditioned and unconditioned stimuli, and when a conditioned stimulus is unreinforced in extinction or differential training, inhibition is supposed to accrue. Kurtsin (1968) lists six methods that Russian scientists use to produce experimental neurosis. Three of these clearly oppose excitation and inhibition in the Pavlovian sense.

1 The duration of a conditioned stimulus for inhibitory reflexes is lengthened.
2 Inhibitory stimuli are frequently presented among excitatory stimuli for conditioned responses.
3 Positive (reinforced) conditioned stimuli are quickly followed by negative (unreinforced) stimuli.

These three methods are all variants of the original Shenger-Krestovnikova procedure. In her method, excitatory and inhibitory processes are opposed because the negative (inhibitory) stimulus resembles the positive one; in the above three procedures the negative stimulus intrudes into times when excitatory processes should be in operation. The modern procedures differ from the original one in that they involve more temporal (GO, NO-GO) than spatial (CIRCLE, ELLIPSE) aspects of the positive and negative conditioned stimulus. In none of these cases are painful stimuli applied; positive stimuli are followed by reinforcement, negative stimuli by nothing.

The other three Russian methods listed by Kurtsin all involve aversive stimulation.

> 4 An ultrastrong sound is used as a conditioned stimulus.
> 5 A strong painful stimulus is employed for forming conditioned responses reinforced with food.
> 6 The meanings of conditioned stimuli are changed so that one that originally signalled food now signals shock, and another that first signalled shock now signals food.

Methods 4 and 5 resemble that of Yerofeeva inasmuch as the conditioned stimuli have aversive properties that remain after the stimuli are associated with food. The procedures capitalize on conflicting tendencies of the painful stimuli to elicit defensive and alimentary responses together, although struggling and emotional arousal could occur on the basis of defensive reactions alone. The sixth method involves features of a procedure developed by Bechterev for conditioning leg flexion with electric shock instead of salivation with food. Pavlov's and Bechterev's methods are used together: conditioned stimulus A, response SALIVATION, reinforcement FOOD; conditioned stimulus B, response LEG FLEXION, reinforcement SHOCK. After the responses are stable, experimental neurosis is obtained when A and B are interchanged.

Miminoshvili (1960) employed a variant of this method with monkeys, noting that these animals, unlike dogs, do not acquire prolonged experimental neuroses on the basis of conditioned alimentary reflexes alone. He employed a Bechterev procedure for establishing a defence reflex, but a Skinnerian operant procedure rather than a Pavlovian one for conditioning the alimentary response. The experimental apparatus is schematized in Figure 5.2.

For the alimentary reflex, the animal was trained to press the lever with food reinforcement in the presence of one tone (S^D) but was not in the presence of another (S^Δ). Then the defence reflex was conditioned, with shock following a red (CS^+) but not a green (CS^-) light. Following this, the meanings of the light stimuli were exchanged, and they were occasionally extended in duration.

> Soon the monkey began to show signs of neurotic state; he ceased taking the food, persistent motor excitability appeared, and he responded to both positive and negative feeding signals with prolonged pressure on the lever or with general motor excitability . . . . The animal reminded one of

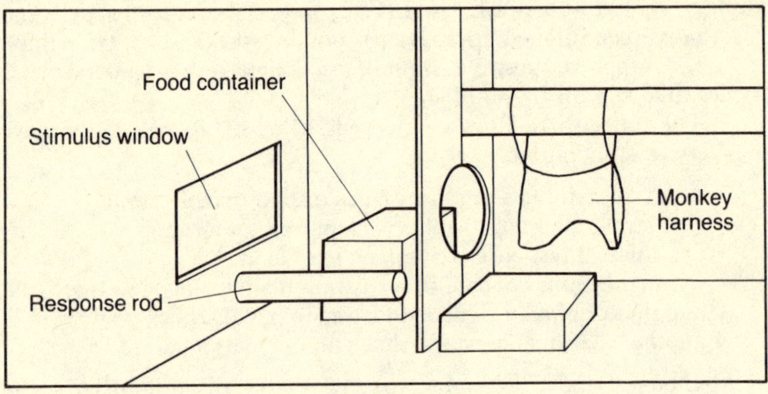

Figure 5.2 Apparatus used by Miminoshvili for the study of experimental neurosis in monkeys. Electrodes for stimulation of the defence reflex are attached to the animal's hind-leg by means of a sock enclosed in a box. In response to the food signal the monkey presses the lever (alimentary motor reflex), which releases food into the food container. Note that the defence reflex is elicited with a non-avoidable shock procedure and that the alimentary motor reflex is a discriminated operant after the manner of Skinner, not an elicited conditioned response after the manner of Pavlov. A photograph of a monkey called Karabus supported in the experimental chamber appears in Miminoshvili, D.I., 'Experimental neurosis in monkeys' in I.A. Usdin (ed.) (1960) *Theoretical and Practical Problems of Medicine and Biology in Experiments on Monkeys*, Oxford: Pergamon Press, 1960, pp. 53–69.

> an automaton. He responded to the positive food signal with pressure on the lever, but not only did he fail to eat the food, he did not even look into the food-container whilst staring fixedly at the lamp – the source of the defence signal.

Later, collision between food and defence reflexes was arranged by having the defence signal coincide with or immediately follow the signal for food. Miminoshvili reports an extremely violent response to this condition with general motor excitability and occasional catalepsy, torpor and drowsiness.

The technique of following a signal for food by a signal for shock also resembles method 3, where a negative signal follows a positive one for food. The more complex method, employing collision of two different reflexes, rather than excitation and inhibition of one, was also used to establish a neurotic reaction in a hamadryas baboon called Volshebnik (Miminoshvili, 1960). This animal showed emotional disturbances in the forms of refusal to eat, erratic lever pressing, screaming and gnawing; and discrimination failures in the paradoxical form of stronger lever pressing to a weak than to a strong stimulus, and the ultraparadoxic form of responses in negative (S^{Δ}) but not positive (S^{D}) stimuli, respectively. Miminoshvili interpreted this form of experimental neurosis as an overstrain of mobility.

Kurtsin (1968, 1976) describes a modern simplification of method 6 with a cat accustomed to feeding on mice.

> One electrode (4 volts) from an electroenergy source was attached to the tail of a mouse, which served as the food stimulus. Then the mouse was put with the cat and the latter being very hungry jumped at once and seized the mouse. But at this very moment . . . the cat received an electric shock. . . . [A] neurotic reaction appeared immediately. The cat expelled the mouse from its mouth and ran away. . . . Such negative reactions to the mouse . . . persisted for many days and weeks . . . . Even a month after the traumatic experience, the sight of a mouse would elicit nervous and avoidance behaviour in the cat. (1968, pp. 81–2)

This modern account of experimental neurosis in a cat may be compared to some behaviour of cats described by Lope de Vega four centuries ago. The passage was drawn to the attention of psychologists by W.A. Bousfield in 1955.

> Saint Idelfonso used to scold me and punish me lots of times. He would sit me on the bare floor and make me eat with the cats of the monastery. These cats were such rascals that they took advantage of my penitence. They drove me mad stealing my choicest morsels. It did no good to chase them away. But I found a way of coping with the beasts in order to enjoy my meals when I was being punished. I put them all in a sack, and on a pitch black night took them out under an arch. First I would cough and then immediately whale the daylights out of the cats. They whined and shrieked like an infernal pipe organ. I would pause for a while and repeat the operation – first a cough and then a thrashing. I finally noticed that even without beating them, the beasts moaned and yelped like the very devil whenever I coughed. I then let them loose. Thereafter, whenever I had to eat off the floor, I would cast a look around. If an animal approached my food, all I had to do was cough, and how that cat did scat!

Although the procedure is similar to that employed by Kurtsin this is not an example of experimental neurosis but a learned adaptive avoidance response. de Vega's procedure has been cited as an instance of Pavlovian classical conditioning, but it is actually a more complex case of instrumental avoidance learning. Saint Idelfonso's pupil's cats could run away; subjects in conditioning experiments cannot. As Kurtsin's cat cannot learn if it has seen the last of electric mice except by eating another, treating all mice with caution is an adaptive, not a neurotic thing to do. The case becomes 'neurotic', however, when there is persistent resistance to extinction of the avoidance response after the danger has passed (see below). This, by implication, was the case with Kurtsin's cat.

In addition to Kurtsin's catalogue of Russian methods for obtaining experimental neurosis, Razran (1971) mentions excessive use of backward conditioning to overstrain the nervous system, and Khananashvili (1976) describes a method that involves conflict of motives. By Khananashvili's brief account, experimental neurosis exists just because normal conditioned responses fail to occur. He describes his method in the following terms.

> A female in heat was placed in the experimental room, where conditioned food reflexes were elaborating in a male. The male, as a rule, came to the female and gave no reactions to conditional stimuli until the end of coitus. Just after that, the conditional food activity recovered without any declination. But, if the female was isolated with a wire-net so that the male could not copulate, an even short-term sojourning of the female in the experimental room disturbed all the conditioned reflexes in the male. A strongly pronounced neurosis was elicited by a long-term placing of the female in the experimental room during conditioning. Thus, the clash of food and sexual motivations conditioned the development of a neurosis only after the unsatisfied sexual excitation.

Thus, the neurosis is *defined* as a failure of normal conditioned responses to appear, and *explained* as the clashing of higher neural impulses. Khananashvili's short report does not describe overt behavioural symptoms of neurosis in male dog, or relate if there was an adverse reaction by the female.The case is probably one of a specific response to a specific situation, and would not meet the generalized criteria proposed by Hebb (1947).

BASED ON INTRAPSYCHIC CONFLICT The most extensive observations of experimental neurosis in cats have been made by Smart (1965), who followed a lead by Masserman (1943). In contrast to the Pavlovian approach to neurosis as a pathophysiology resulting from clashes of excitatory and inhibitory neural impulses, Masserman adopted the psychoanalytic view of neurosis as a conflict of intrapsychic forces. He confined cats in a glass cage and trained them to operate a switch that flashed a light, rang a bell and deposited a pellet of salmon into a food-box. After several months on this routine the animals were given either a mild air-blast across the snout or electric shock through the paws as they obtained the salmon. With from two to seven such experiences, Masserman (1950) claims, the cats showed aberrant behaviours resembling human symptoms of neurosis.

> Neurotic animals exhibited a rapid heart, full pulse, catchy breathing, raised blood pressure, sweating, trembling, erection of hair . . . . They . . . became 'irrationally' fearful not only of physically harmless lights or sounds but also of closed spaces, air currents, vibrations, caged mice and food itself. The animals developed gastro-intestinal disorders, recurrent asthma, persistent salivation or diuresis, sexual impotence, epileptiform seizures or muscular rigidities resembling those in human hysteria or catatonia.

Several additional behavioural disturbances in cats and other animals are mentioned by Masserman and his colleagues, e.g. Masserman and Siever (1944), but as these are often anecdotal, Smart (1965) conducted a careful experiment to test them. He employed thirty cats which were trained to approach and open a food-box on signal. After approach training, dominance tests were made in which paired animals were placed in the apparatus to see which would take food first when the approach signal flashed and sounded. Finally, conflict sessions like Masserman's were conducted. However, at this point the cats were divided into three equal groups of ten, a *consummatory* group, a *pre-consummatory* group and a *shock alone* group. The ten cats in the *consummatory* group were shocked after they began eating; the ten cats in the *pre-consummatory* group were shocked after they had opened the food box but before they began eating; and the cats in the *shock alone* group were shocked with at least 30 seconds between the shock and consummatory or pre-consummatory behaviour. This last group was employed to eliminate the element of approach–avoidance conflict, taken by Masserman to be the basis of experimental neurosis. Shock sessions occupied four days for each group, in which eight shocks occurred randomly during forty experimental trials per day. The cats were observed before and after shock and rated according to a complex set of sixteen scales.

Table 5.1 shows the number of cats (out of thirty) whose behaviour changed due to the shock experience. All animals were rated non-neurotic before shock, but most of them exhibited neurotic reactions by the end of the experiment. No tests were made for changes in inter-animal conflicts. Statistical tests for

Table 5.1 Most frequent neurotic reactions of thirty cats as a result of electric shock in the experimental apparatus. It made no difference whether shock was associated with feeding (conflict) or not

Type of response	*Number of cats*
Loss of dominance	30
Neurotic food avoidance	28
Neurotic switch avoidance	28
Neurotic motor disturbance	27
Reaction to food box	25
Fear of constriction	24
Attraction to apparatus	22
Neurotic hypersensitivity	21
Autonomic changes	21
Escape behaviour	20

After Smart, R.G. (1965) 'Conflict and conditioned aversive stimuli in the development of experimental neurosis', *Canadian Journal of Psychology*, 19, 208–22.

differences in the number of animals per group exhibiting neurotic symptoms showed no differences according to experimental treatment, thus eliminating conflict as a necessary originating condition for experimental neurosis.

More recently, Dmitruk (1974) has argued that even the aversive stimuli employed by Smart are unnecessary for establishing behavioural disorders in experimental cats, because symptoms of experimental neurosis are elicited merely by placing the animals in confinement. Examples of 'cage stereotypies' in other animals are described in Chapter 3. They are not normally equated with generalized neurosis but are attributed to the specific experimental circumstance of confinement.

BASED ON FAULTY LEARNING Smart's (1965) findings agree with a theory of experimental neurosis proposed by Wolpe (1952). Wolpe argues that experimental neuroses are learned responses to the experimental situations, and that they therefore should have three features in common with other learned behaviour (Wolpe, 1967).

1 The neurotic behaviour must be closely similar to that

evoked in the precipitating situation.
2 The neurotic responses must be under the control of stimuli that were present in the precipitating situation.
3 The neurotic responses must be at greatest intensity when the organism is exposed to stimuli most like those to which the behaviour was originally conditioned.

Wolpe's (1952) own experiments with cats showed these features, and on the basis of his findings with animals Wolpe devised the therapeutic technique of counterconditioning by reciprocal inhibition that is now a common treatment for human phobias. This is the treatment described in Chapter 1 as it was applied to the sheepdog, Higgins. It involves strengthening relaxation under different conditions from those that evoke the neurosis, and gradually adjusting the conditions towards the neurosis-provoking situation, all the time keeping relaxation prepotent. This is a more complex form of deconditioning treatment than an earlier one applied to animals by Merrill (1945).

Merrill cured two great Danes of sheep-killing by repeatedly exposing them to penned lambs that the dogs could not reach. At first the dogs were greatly excited, but gradually the excitement vanished, and after a few weeks when the lambs were freed, the dogs ignored them. A Dalmatian was cured of killing chickens in a similar way. These dogs would not be called neurotic, but neither would Wolpe's experimental animals if Hebb's criteria were applied, for Hebb's criterion of neurosis as chronic, generalized emotion is in direct conflict with Wolpe's claim that neurosis is situation-specific. However, Hebb's criteria are based on symptomatology; Wolpe's claim stems from his analysis of neurosis by way of originating conditions.

Another learning-theory approach to experimental neurosis relates it to learned helplessness. Thomas and DeWald (1977) report an experiment with cats based on the Shenger-Krestovnikova method of producing experimental neurosis. The cats were required to perform a behavioural chain in which they pressed one lever (R_1) to expose two others, a bright one and a dark one. Then, pressing the bright lever (R_2) was reinforced with food but pressing the dark one was not. After a reliable discrimination had been established, such that the cats performed the chain $R_1 - S^D$ (light) $- R_2 - S^R$ (food), the dark lever was progressively brightened to the level of the original bright lever, whereupon the animals first displayed agitated aggression but eventually stopped pressing the lever that initiated trials (R_1) and 'sat or lay immobile with their shoulders rigidly hunched in a distinctly depressive posture that is characteristic of experimental neurosis' (Thomas and DeWald,

1977, p. 222). This description is quite unlike the squealing, wriggling and barking behaviour that Pavlov (1927) describes for Shenger-Krestovnikova's dog, but does resemble some of the behaviour seen in experiments on extinction and extinction-induced aggression.

However, manifestations of experimental neuroses can take many different forms, including the various discrimination failures described above. These forms are not necessarily those of human neurotic sub-categories, but an analogue of chronic anxiety can be seen in the persistence of avoidance behaviour during extinction procedures. Avoidance behaviour is behaviour that avoids or postpones aversive events. A typical experimental procedure is that of Sidman (1966) in which brief electric shocks recur at a specified shock–shock interval (SS-5 seconds, say) until a specified response occurs. Such a response initiates a response–shock interval (RS-30 seconds, say) that is free of shock. Further responses in the response–shock interval continue to postpone shocks, so with a

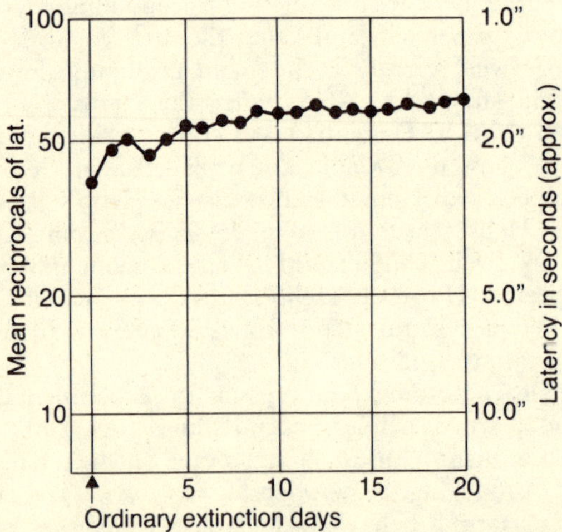

Figure 5.3 Mean reciprocal latencies, and approximate actual latencies, of thirteen dogs over ten non-shock extinction trials on each of ten successive days. Latencies are times for jumping from one compartment to another after signal onset. At criterion of shock-avoidance learning (arrow), mean latency was 2.7 seconds. After 200 shock-free extinction trials, mean latency had declined to approximately 1.6 seconds even though jumping was no longer functional. Note that the right-hand latency scale is inverted. (From Solomon, R.L., Kamin, L.J., and Wynne, L.C. (1953) 'Traumatic avoidance learning: The outcome of several extinction procedures with dogs', *Journal of Abnormal and Social Psychology*, 48, 291–302.)

steady rate of responding an animal can avoid shocks altogether. When the avoidance response is a barpress it is difficult for rats to learn this procedure because instead of repeatedly pressing the bar they 'freeze' with it held depressed, but running in a wheel and shuttling in a box are relatively easily learned by rats and dogs with varieties of this procedure.

What happens to avoidance behaviour when shocks are no longer programmed? Kurtsin, as mentioned above, gave the answer 'experimental neurosis', but frequently avoidance extinction occurs quite rapidly. Sometimes, however, a well-trained animal may continue responding for hundreds of trials unnecessarily. Figure 5.3 illustrates a finding of Solomon, Kamin and Wynne (1953) with dogs trained to avoid traumatic electric shocks by jumping from one compartment to another in a shuttlebox. Potential shocks were signalled by lowering the barrier and sounding a buzzer or changing the lights 10 seconds before each shock was due, and the animals learned to avoid shocks by jumping the barrier in the time between the signal and possible shock onset. Figure 5.3 shows the mean response latency of thirteen dogs on twenty extinction days with ten shock-free trials per day. After 200 such trials the dogs still jumped promptly on signals, with latencies that were shorter at the end than at the beginning of extinction, although there was no penalty for not jumping.

In an effort to hasten extinction, the experimenters introduced shock for three seconds in the compartment that the dogs jumped into. In a pilot experiment with two dogs, one dog, which had received only eleven shocks in training, jumped for 490 trials in extinction, at which point shock for jumping was introduced, whereupon:

> The dog became more upset, and at subsequent presentations of the CS jumped more vigorously . . . . After 100 additional trials under this shock-extinction procedure, the dog was still jumping regularly into shock and gave no signs of extinguishing. As he jumped on each trial, he gave a short anticipatory yip which turned into a yelp when he landed on the electrified grid.

Of the thirteen dogs in the main shock-extinction experiment, ten continued jumping for all 200 shock-free trials, with decreased latencies and greater vigor than before. Solomon and Wynne (1954) appeal to partial irreversibility of conditioned anxiety to explain these findings, but the appeal is *ad hoc* and is not supported by independent evidence.

Experimental neurosis evaluated

Plainly, an experimental analogue of a phenomenon must have a clear and objective description of the model to begin with. In the case of neurosis as a disease state this has never been achieved, although it may have been with some sub-categories of clinical pictures. Hebb (1947) long ago criticized experimental neurosis as unsatisfactory evidence of abnormal emotional states induced in experimental animals. He listed four claimed indices of experimental neurosis that he considered to be not neurotic but normal responses to the traumatic circumstances of the experiment: unusual test behaviour, like somnolence and refusal to eat; attacks on the apparatus and refusal to enter the experimental room; failure to eat outside the apparatus; and disturbances outside the test situation. Only discrimination failure was acknowledged as real evidence of an abnormal state in the experimental animals but, as Hebb notes, this is not analogous to an undesirable, emotional, persistent and generalized state of human neurosis.

Inasmuch as the revised *Diagnostic and Statistical Manual* (*DSM-III*) has eliminated neurosis as a disease entity in favour of more specific clinical pictures, it appears that the experimental evidence has antedated, and possibly influenced, psychiatric clinical experience. In that respect Hinde's (1962) 20-year-old contention that the relevance of animal studies to the problem of human neurosis lies in the study of animal responses to stress might be a better experimental strategy than that of seeking equivalences between animal and human diagnostic categories.

On the other hand, an analysis of neuroses based on conventional diagnostics is offered by Blanchard and Hersen (1976), who divide neurotic disorders into syndromes maintained by anxiety reduction and syndromes maintained by secondary gain. Phobias, obsessions, compulsions and some depressions are included in the first group; conversion hysteria belongs in the second. This division does not correspond to the one suggested earlier in this chapter, wherein anxiety neurosis and hysteria were divided from obsessional and phobic neuroses on the basis of autonomic-response criteria, but it does relate to the classical differentiation of anxiety states from conversion hysterias on account of 'la belle indifference' shown by the latter kind of patient. Blanchard and Hersen expect symptom substitution in neuroses maintained by secondary gain, but not in neuroses that depend on anxiety reduction. In that case, animal experimental neurosis of the hysterical type could be any kind of superstitious or anomalous operant behaviour, like the misdemeanours described in Chapter 4, and experimental neurosis maintained by anxiety reduction would be the resistance to avoidance extinction described in the preceding section.

Psychophysiological disorders

Psychosomatics and the specificity question

SPECIFICITY AND PSYCHOANALYSIS Psychophysiological disorders, also called psychosomatic illnesses, are 'characterized by physical symptoms that are caused by emotional factors and involve a single organ system, usually under autonomic nervous system innervation'. In the American Psychiatric Association diagnostic manual they are classified according to the physical system affected, in particular as disorders of the skin, of the musculoskeletal system, of the respiratory system, of the cardiovascular system, of the haemic and lymphatic system, of the gastro-intestinal system, of the genito-urinary system and of the endocrine system. Also, such afflictions as neurodermatitis, tics and muscle cramps, bronchial asthma, rheumatoid arthritis, hypertension, peptic ulcer, amenorrhea, diabetes and certain tumours and growths are often attributable to psychological origins. Wright (1977) proposes to classify as psychosomatic psychological disorders caused by physical factors as well as physical disorders caused by psychological factors, but Lachman (1972) classifies these separately as somatopsychic disturbances. The present section is concerned only with physical disorders of psychological origin, or with neurogenic disorders, as Startsev (1976) calls them.

It is widely acknowledged that neurogenic disorders are the result of emotional stress. Selye (1956) defines stress as 'the nonspecific response of the body to any demand' and reports several bodily responses to stressful situations, including adrenal enlargement, shrinkage of lymphatic organs, gastro-intestinal ulceration, hormonal disturbance and general loss of bodyweight. He differentiates this general adaptation syndrome (G.A.S.) from a local adaptation syndrome (L.A.S.) that is the response of a particular bodily organ to assault.

In neurogenic disorders the mechanisms relating nonspecific stress to specific physical systems are poorly understood. One form of 'specificity hypothesis' relates characterological and predispositional factors to particular psychosomatic manifestations – what Engel (1981) calls individual-response specificity. In this light a psychoanalytic group (Alexander, French and Pollock, 1968) attempted to predict from carefully edited psychoanalytic interview protocols alone which of seven diseases – bronchial asthma, rheumatoid arthritis, ulcerative colitis, essential hypertension, neurodermatitis, tyrotoxicosis and duodenal peptic ulcer – particular patients were afflicted with. On the assumption that each illness is associated with a particular type of 'onset situation' and a particular 'psychodynamic constellation', (e.g. 'The central dynamic feature in duodenal peptic ulcer is the frustration of dependent desires originally oral in character', and 'In cases of neurodermatitis we find a complex configuration between exhibitionism, guilt and masochism, combined with a deep-seated desire to receive physical expression of love from others . . . . The disease as a rule, is precipitated after the patient achieves some form of exhibitionistic victory.') the analysts were significantly better than a group of internists in matching psychological data with actual physical illnesses in most cases. However, correspondence between the psychological and physical data was poor in many cases, and the analysts admitted that both the psychodynamic constellations and the onset situations are common in individuals who do not suffer from psychosomatic illnesses. Individual constitutional organ weaknesses were postulated to differentiate patients and non-patients, but evidence of differences was not presented.

An exact correspondence between a psychosomatic tic and an originating circumstance appears in a case with psychoanalytic overtones described by Gerard (1948). Charles, a nine-year-old boy was referred for treatment on account of his withdrawal through fear from all forms of aggressive activities, including sports, which were compulsory at his school. 'If forced to play, he fumbled balls, ran very poorly and trembled noticeably. He preferred to play with girls and would dress often in his sister's clothes and parade before a mirror.' In addition, he had a tic wherein he 'would suddenly ball his right hand into a fist over his genitals and quickly run the fist up the fly of his pants as if hitching them up. At the same time, he jerked his head backward and to the left' (Gerard, 1948).

Gerard (1948) proceeds with a detailed account of the life history of the boy, which includes the following events.

> At 6 a traumatic event occurred. The father, going into his bathroom one afternoon, caught Charles naked in front of a long mirror masturbating. In a rage he slapped Charles . . . [and ordered him] to bed without his dinner . . . . In the morning the father admonished him again and told him that if he ever masturbated again, the doctor would have to cut the foreskin. While the father was talking Charles put his hand over his pants fly and the father raised his hand to slap him. Charles jerked his head back and looked so frightened

> that the father restrained himself from hitting him. It was that evening that the mother noticed the tic when the family was eating dinner. (Gerard, 1948, p. 481)

The relationship between symptomatology and origination in this instance are dramatic and clear, but such examples of specificity of symptoms with respect to an originating situation are relatively rare in psychosomatic medicine, and the case of Charles might alternatively be seen as an hysterical reaction. It is a case of faulty learning in which a reflexive defensive avoidance behaviour fails to extinguish, but re-occurs in similar situations in which no immediate danger exists.

SPECIFICITY AND EXPERIMENTAL NEUROSIS With animal experimental models of psychosomatic illnesses, Bowden (1976) set out some alternatives to the specificity question in a discussion of experimental neurosis created by Pavlovian techniques, noting in particular the five patterns of pathology of the higher nervous system listed above. These are breakdowns of conditioning in dogs already trained to make stable conditioned salivary responses, and pertain particularly to disturbances in discrimination.

In addition to these, however, much additional behaviour also occurs in experimental neurosis, including disturbances of physiological functions. These latter disturbances are not 'symptom-specific' in the sense that particular stressors produce particular pathological patterns in all dogs (or at least those dogs with weak nervous systems that break down under stress) but appear to relate more to the prior history of an animal. It is this appearance that stimulated Startsev (1976) to propose a conditioning theory of specificity such that the particular system affected by stress is the one that is activated on the occasions of stressful experiences. As he says:

> In essence, the conditioned and unconditioned stimuli, which normally stimulate a certain physiological system [e.g. gastrointestinal, motor, sexual, carbohydrate metabolism], are transformed into conditioned signals of a defensive or other pathological dominant. They lose, to a certain degree, their former physiological significance and become the chief pathogenetic factor. (Startsev, 1976, p. 12)

By defensive pathological dominant is meant a powerful defensive drive that chronically dominates the physiological and psychological status of an organism. Startsev establishes it by restraining an animal in an apparatus in which it can do little else but press a lever for food. So confined, a hamadryas baboon will struggle, screech, tremble, claw the stand and exhibit 'a constant state of emotional excitement, except for transient periods when he [is] limp from exhaustion'. According to Startsev, whatever physiological system is operative at the time such a defensive dominant is aroused becomes a conditioned signal for the defensive dominant and thereby acquires a functional, or neurogenic, pathological character.

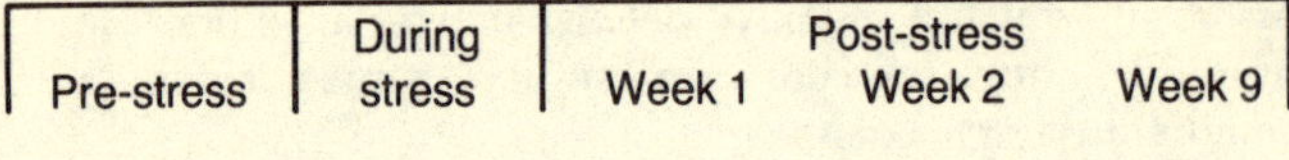

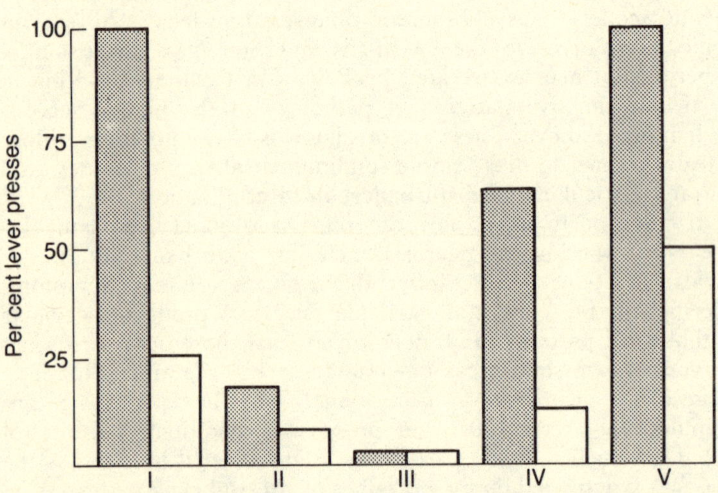

Figure 5.4 Percentages of lever presses during S^D (solid bars) and S^Δ (open bars) stimulus presentations before, during and after five experimental sessions in which immobilization stress followed home cage feeding. After Startsev, V.G. (1976) *Primate Models of Human Neurogenic Disorders*, Hillsdale, N.J., Erlbaum.

Figure 5.4 shows a breakdown of the stimulus control of lever responses reinforced by food in the hamadryas baboon, Larn, after the animal had suffered repeated combinations of normal feeding in the home cage with subsequent 5-hour fixations in the restraint stand. The animal was first trained with a discrete trial procedure with a stereotyped series of high and low tones and white and red lights in the order shown in Table 5.2, where a lever response was reinforced with food in the presence of the positive stimuli (S^D) and

Table 5.2 Stereotyped sequence of stimulus presentations used by Startsev in the establishment of conditioned alimentary motor responses (lever presses) by baboons. The reinforcer was food. Once discriminated lever pressing in S^D was established immobilization stress in the conditioning apparatus after home cage feeding was introduced

Stimulus order	*Reinforced discriminitive stimulus (S^D)*	*Unreinforced differential stimulus (S^Δ)*
1	Low tone	
2		High tone
3	Low tone	
4		High tone
5	White light	
6		Red light
7	White light	
8		Red light
9	Low tone	
10		High tone
11	White light	
12		Red light
13	Low tone	
14	White light	

After Startsev, V.G. (1976) *Primate Models of Human Neurogenic Disorders*, Hillsdale, N.J., Erlbaum.

was not reinforced in the presence of the differential stimuli (S^Δ). The left hand pair of bars (I) of Figure 5.4 show that the animal responded appropriately to the combined positive (S^D) and seldom to the differential stimuli (S^Δ) before immobilization stress began. Pair II shows the effect on lever press rates when experimental sessions were combined with immobilization stress, and pairs III, IV and V show residual effects 1, 2 and 9 weeks after the pairing of immobilization with home cage feeding was discontinued. With recovery, not only lever press frequency but also its duration and intensity increased.

In addition to disruption of lever pulling for food, food intake was interrupted, and even when eating returned to normal it was followed by vomiting and inhibition of gastric juices. Such experimental neurogenic gastric achylia (inhibition of gastric juices or production of alkaline juices with little or no digestive potential) was produced in nine out of nine hamadryas baboons subjected to repeated pairings of eating with immobilization stress. Not one of ten control animals that were immobilized during the fasting state showed more than transient hypoacidity after stress.

Also, pretreatment with haloperidol (that tranquilized subjects during immobilization periods) attenuated the alimentary disorders caused by pairing eating with stress. These data, too, Startsev (1976) claims, support his theory that the particular neurogenic disorder caused by stress occurs in the physical system that either spontaneously or experimentally becomes transformed into a conditioned interoceptive signal for defensive excitation.

The nine baboons that developed neurogenic gastric achylia as a result of immobilization stress also exhibited precancerous stomach lesions. Startsev first observed such lesions close to fistulas implanted for the collection of gastric secretions, but found that they also occurred in the animals without implanted fistulas. Similar results in dogs subjected to neurogenic stress by Pavlovian methods were reported by Petrova in 1946.

Startsev also describes another of Petrova's cancer experiments in which for two years:

> Coal tar was applied daily to a 10 × 10 cm area on the back of nine dogs (1–2 years old at beginning of experiment). Four of the dogs, the experiment animals, were subjected to experimental neurosis after the first year, whereas the other five, the control animals, did not undergo neurogenic stress. The stressed dogs showed papillomatosis at the site of chronic irritation. In addition, one of them developed a tumor near the parotid gland and another developed a sarcoma of the urinary bladder with metastasis to the kidneys, liver, intestine, and spleen. The control dogs developed papillomata at the site of the coal tar application, but the lesions were only transient, subsequently disappearing without a trace. (Startsev, 1976, p. 102)

Altogether, Startsev describes psychogenic heart disease, sexual disturbance, hyperkinesis and paralysis as well as gastric disorders in experimentally stressed baboons. Chronic hyperglycemia was not produced in baboons by similar methods, although it did occur with rhesus monkeys.

The contingent model for altering physiological states

Brady and Harris (1977) distinguish concurrent and contingent models of interactions between physiological states and environmental events. Concurrent models are those described above, where physiological disturbances are found to accompany environmental stress, and where the analytical problem is to discover how particular physiological systems are affected in individual patients. The answer appears to lie as much in relationships between environmental events as it does in characteristics of individuals, and the same is true of contingency models of psychosomatics, except that the emphasis is on operant instead of on classical conditioning. In the contingency case, attention focuses on therapy rather than on aetiology, and the question is the extent to which autonomic systems, like skeletal systems, respond to operant contingencies. Thus, whereas analyses of the acquisition of psychophysiological disorders typically proceed in classical conditioning terms, the techniques for the elimination of such disorders derive from operant conditioning concepts and procedures. These techniques are known collectively as biofeedback.

Although there are theoretical reservations about the mechanisms by which autonomic responses are operantly conditioned – directly in response to environmental contingencies or indirectly via skeletal mediating responses – the application of biofeedback therapies to such psychophysiologic disorders as essential hypertension, cardiac arrhythmia and peptic ulcer has not been retarded by them. In fact, such techniques, along with behaviour therapy and behaviour modification, represent powerful contributions of animal models to clinical psychology and psychiatry. These models, it is worth noting in passing, originated in studies of normal animal behaviour and learning that had no relationship either to psychiatry or to psychosomatic medicine.

Biofeedback has passed from the realm of experimental animal models of psychiatry into that of established clinical practice. In this connection, Engel (1981) cites two ways in which this practice is now employed – one to control specific physiological manifestations and the other to control non-specific actions, particularly muscular tension. From his experience with these practices Engel derives two salient practical principles: the reinforcing property of success; and the specificity of modified responses. The first of these principles points to the powerful effect on patients of mastery over their own autonomic functions, and the second illustrates the point that effects of biofeedback are specific to the response under training – they do not generalize even to closely related functions. As examples, Engel cites the cases of monkeys whose heartrate changes were conditioned without affecting blood pressure and, of a human female who was trained to control her heartrate and intracardiac conduction patterns independently.

6 Animals and addictions

Natural animal-drug interactions

Humans use drugs for therapeutic purposes, and in this respect they have followed animals in the discovery of natural remedies and poisons. More of a mixed blessing is the human recreational use of drugs, and here also there are animal–human parallels. R.K. Siegel (1979) compared self-administration of tobacco, alcohol, hallucinogens, stimulants, cannabis, opium and coca by 109 different animal species and 144 hunting and gathering human cultures and found the results shown in Table 6.1. From these Siegel proceeded to document several examples of natural animal addictions, including alcohol, tobacco, narcotics and catnip, defining addiction in the Latin sense of fondness or passion.

Addiction is sometimes defined more particularly as 'habituated to' or 'dependent upon'. This may be called 'true addiction', implying a passion taken to the point of harm. Leyhausan (1973) cites the catnip response as illustrative of such an animal passion.

> When the Japanese investigators started to use [an extract of *Actinidia polygama*] in some concentration on the large cats of Osaka Zoo, a very peculiar phenomenon emerged: after a few experiments these cats became so eager that the moment they saw the experimenter appear they left everything, including food, sexual intercourse, or whatever it was, and just ran up to the bars and waited for this smell. As soon as they got it on a tuft of cotton wool, they reacted very intensely (by vigorously rubbing chin and cheeks against it). Yet, unlike animals performing a normal catnip response, they all rolled on their backs in the end and stayed there for some time in complete ecstacy. (Leyhausan, 1973, p. 62)

He speculates, 'I believe the animals become truly addicted to this smell, because they continue to react although it affected their

Table 6.1 Comparative distribution of major groups of psychoactive plant drugs used by cultural groups of man (N = 144) and species of animals (N = 109) in natural habitats

Man		*Animal*	
Plant drugs	*Number of cultures*	*Plant drugs*	*Number of species*
Tobacco	57	Alcohol	28
Alcohol	52	Tobacco	20
Hallucinogens	40	Hallucinogens	19
Stimulants	15	Cannabis	18
Cannabis	7	Stimulants	9
Opium	3	Opium	9
Coca	1	Coca	6

The human data are adapted from Blum (1969.) Correlation with animal data is $r = .90$, $p < .01$. From Siegel, R.K. (1979) 'Natural animal addictions; an ethological perspective', in Keehn, J.D. (ed.), *Psychopathology in Animals: Research and Clinical Implications*, New York, Academic Press, pp. 29–60, reprinted by permission.

sense of smell and in the end damaged the brain, as my Japanese hosts told me.'

Cats in particular respond to catnip, but the tendency to chew and gnaw avidly all manner of substances is common in a variety of animals. When the substances are aversive to humans the phenomenon is called abnormal appetite or pica, although the appetite can be normal in the sense of satisfying a body deficiency. An example of pica with humans is coprophagia, the eating of faeces, an indulgence of some psychiatric hospital inmates that also occurs in monkeys kept in close confinement. A list of morbid animal appetites prepared by Stainton (1941) long ago includes:

1 licking disease in cattle;
2 wool eating in sheep;
3 sand eating in horses;
4 plucking of own feathers by birds;
5 the aberration of the pregnant animal, which shows inordinate desire for material (food or otherwise) which would ordinarily be avoided;

> 6 the eating of cinders, coal, wood and horse manure by *young* dogs;
> 7 the sudden preference for filth, dirty water and human excreta above good food by the *old* dog;
> 8 the dog suffering from rabies who, during the premonitory stage, will avoid favourite food and begin to bite and gnaw anything within reach and swallow foreign bodies such as earth, straw, glass, rags, and even its own faeces and urine.

Stainton also asserts that nearly all animals will acquire the taste for alcohol, and in reference to alcoholism gives several anecdotal examples including the following.

> My father, who was a veterinary surgeon in the old horse-and-trap days, owned a pony called Moses, and this pony was generally driven by the assistant on his rounds. Apparently this assistant had been in the habit of calling at a wayside pub, which meant driving into the usual adjacent parking enclosure, and as he was not a lonely drinker he always brought out a pint of beer for the pony. One day my father had occasion to use the pony himself and, driving with a loose rein, suddenly found himself in the precincts of this hostelry, where Moses had arrived for his drink. Explanations followed, but it was a puzzled and disappointed pony that was coaxed back to the straight and narrow road. (Stainton, 1941, p. 25)

But this is nothing compared to the beer and spirit drinking habits of circus and laboratory elephants (Siegel and Brodie, 1984). In the wild, the nearest approach found to human alcohol consumption is the case of elephants tippling on fermented mango fruit. Illustrations such as these (see R.K. Siegel, 1979, for an extensive survey of animal involvement with natural psychoactive plants) confirm that intoxication is not a uniquely human situation, but they are not informative about the acquisition and maintenance of alcohol seeking and consuming behaviours that have deleterious effects on certain humans. For answers to these problems controlled laboratory studies are essential.

One approach to this endeavour is to establish criteria of alcoholism (e.g. Davies, 1976) and seek out animal models that meet them. In this way some six animal models have been investigated – a nutritional model, a genetic model, a neurophysiological model, a pharmacological model, a reinforcement model, and an adjunctive behaviour model – but none is entirely satisfactory. This may be because the models are faulty, because the criteria are inadequate, and/or because the search for an animal model *of* alcoholism is an erroneous strategy. An alternative strategy is to seek animal models *in* alcoholism after the manner described below.

Animal models in alcoholism

Cicero (1980) gives three answers to the question of whether an animal model of alcoholism is possible: no, maybe and yes, depending on how alcoholism is defined. Given that an alcoholic is 'any individual whose repeated and continued use of alcohol interferes with the efficient performance of his work' (Trice, 1970), Cicero's negative answer must be accepted. However if Rodgers (1967) is right, that 'Alcoholism, by conventional definition, is the development of a nonspecific physiological or sociological pathology that is aggravated by, and presumed to be the consequence of, a chronic history of alcohol ingestion', then the 'yes' answer could be accepted because he induced physiological and behavioural deficits in mice by feeding them alcohol for about a year. The 'maybe' answer applies when the question is phrased like the title of this section. In that case, animals may serve in the study of factors affecting alcohol preferences (genetic model), in the study of methods for inducing high alcohol intake (reinforcement model, polydipsia model) and for the comparison of motor, perceptual, neural and other effects of alcohol between animal species and man.

With respect to animal models in alcoholism that are feasible, Cicero proposes criteria for four measurable aspects.

1 *Self-administration* in pharmacologically measurable amounts by the oral route such that the animal overcomes obstacles to obtain it.
2 *Tolerance* should develop after chronic self-administration.
3 *Signs, symptoms, and reactions of withdrawal behaviour* should appear upon withdrawal of alcohol from the animal.
4 *Biomedical complications*, like liver and brain damage, should appear.

Lester and Freed (1973) have a similar but differently worded set of criteria, with the additional requirement that after abstinence there should be reacquisition of drinking to intoxication, and reproducibility of the alcoholic process. Here, I shall discuss matters of self-administration, tolerance and withdrawal with respect to a comparative psychology of alcohol dependence. Biochemical complications in liver, heart, brain and other body organs reported in numerous laboratory studies employing a variety of experimental animals are reviewed by Wallgren and Barry (1970), Eriksson, Sinclair and Kiianmaa (1980) and several other sources. They bear on the deleterious effects of alcohol on the body, but not on the acquisition and maintenance of behavioural responses to alcohol, with which comparative psychology and psychiatry are concerned. The question of readdiction is addressed below in the context of opiate addiction, where examples of conditioned withdrawal syndromes responsible for readdiction are more plentiful than in the case of alcohol, both with animal and with human subjects.

Self-administration

Laboratory studies of alcohol ingestion by animals employ intravenous, intragastric and oral routes of administration. Of these, the intravenous route delivers alcohol most rapidly to the brain. The intragastric route avoids problems of smell and taste avoidance often found in animals, but the oral route is the one used by voluntary human imbibers for whom the animals are supposed to be models.

Voluntary self-administration of alcohol by rats, mice, monkeys and other laboratory animals by this route is achieved by means of genetic selection, reinforcement *of* drinking, reinforcement *by* drinking, and schedule-induced drinking. The first of these involves selection from two or three bottle alternatives, the second makes feeding contingent on alcohol-drinking, the third employs alcohol to reinforce other behaviour, and the fourth is a side-effect of intermittent feeding.

With respect to genetic selection, Rodgers (1967) began an award-winning essay by invoking an Old English expression 'Drunk as a mouse.' A similar expression 'Drunk as a lord' might have led him to the Mughal conquerors of India or the House of Medici in Italy, but the genealogies of the Mughal Jehangia and the Medici Gian Giovanni (Acton, 1932) would have been less helpful than that of the C57BL/Crgl alcohol-preferring strain of mouse that Rodgers employed. A group of this strain was maintained under controlled laboratory conditions for a year with standard diet and free choice of 10 per cent alcohol or water to drink, after which time they appeared less healthy than members of a control group that drank only water over the same period of time. This does not provide an animal model of alcohol*ism*, as Rodgers (1967) cautiously proposed, but it did encourage others to attempt selective breeding programmes for alcohol preferring and non-preferring animals. Among these, Eriksson (1980) concludes that although alcoholism is not heritable as such, voluntary consumption quantities, enzyme systems regulating alcohol metabolism, susceptibility to intoxication and sleeping times under the influence of alcohol, do have genetic bases in rats. He acknowledges, however, that 'the question of their respective significance in drinking behaviour remains unanswered', although the most probable contribution of genetics to alcohol ingestion is via the avoidance rather than the acceptance of the substance.

Oral consumption of alcohol by rats has been achieved experimentally with both extrinsic and intrinsic reinforcement. With extrinsic reinforcement, Keehn (1969) obtained blood-alcohol levels up to 200 milligrams alcohol per 100 millilitres of blood in hungry rats that were reinforced with food for licking alcohol in half-hour sessions twice a day. After 2-months' experience of this regimen the animals had consumed a considerable amount of alcohol at intoxicating levels but on a choice test they chose plain over alcohol-treated saccharin. With intrinsic reinforcement, rats of the drinking strain developed by Eriksson (1980) on the basis of two-bottle choice preference, barpressed for alcohol in operant conditioning boxes (Sinclair, 1974). These rats appeared not to consume intoxicating quantities, but insofar as they drank more alcohol the less the effort required to obtain it, they behaved like human alcoholics (Mello, 1972). Meisch (1980) describes similar studies in which rats and monkeys, first trained while hungry to barpress with alcohol as the reinforcer, continue to work for the beverage when no longer food deprived.

These methods induce self-administration of alcohol by laboratory animals through the oral route, but the amounts ingested are not especially excessive. Rats drink large amounts of alcohol if it is suitably sweetened, but the method most particularly suitable for inducing sustained high intakes of alcohol by rats, monkeys and mice is based on an accidental discovery that rats intermittently reinforced with dry food pellets for barpressing develop the habit of drinking after every pellet and rapidly consume several times their normal daily water intakes. Lester (1961) used this method with alcohol, and since then several groups have employed it to induce excessive alcohol consumption by laboratory animals. Alcohol drinking is not directly reinforced by this method but is a byproduct of the scheduling of food reinforcement at intervals one or two minutes apart. Such schedule-induced drinking is thought to resemble human alcoholism insofar as it involves excessive consumption and withdrawal symptoms when the availability of alcohol ceases.

Schedule-induced drinking may turn initially unpalatable alcohol into a positive reinforcer. However, rats that choose certain alcohol concentrations over water in their home cages may reverse this choice for schedule-induced drinking (Keehn and Coulson, 1975), and even physically dependent rats starved almost to the point of death favour a sweet saccharin over a weak alcohol solution (Samson and Falk, 1974). In any event it is not difficult to make laboratory animals self-administer alcohol in chronically intoxicating quantities by the intermittent feeding technique, although the extent of self-administration depends on available alternatives.

Lester and Freed (1973) deny that the schedule-induction method gives a suitable basis for an animal analogue of human alcoholism, arguing that the alcoholic drinks from strong inner motivation, not because of external manipulations as in the case of the rat model. In this they express a traditional bias that may be wrong, for after an extensive review of numerous behavioural studies of drinking by human alcoholics one author concluded that:

> The data summarized in this review represent a very different picture of the effects of alcohol on the alcoholic than has been derived from anecdotal accounts based on the self reports of alcoholic subjects during sobriety. The voluminous psychiatric literature on alcoholism has tended to present 'the alcoholic' as an impulsive hedonist who drinks to dissolve his anxiety and achieve a diffuse sense of omnipotence. Moreover, the alcoholic has been described as a person possessed by a demonic 'craving' for alcohol and that 'the first drink' would trigger an uninterrupted sequence of compulsive drinking until he achieved a 'state of oblivion' or the alcohol supply was exhausted. Empirical observations have not supported these prevailing notions about alcoholism and thereby re-emphasize the need for systematic investigation of the basic phenomenology of this complex behaviour disorder. (Mello, 1972, p. 280)

Mello's 'very different picture' is not a matter of a different interpretation of alcohol abuse, but a different description of what there is to be interpreted. Systematic investigations with humans have shown that some traditional criteria of animal models in alcoholism may not need to be met. Four such non-criteria are:

1 craving and loss of control;
2 anxiety reduction after drinking;
3 alcohol preference; and
4 unremitting effort for a drink.

Human alcohol abusers drink for many different reasons, some of which are not available to lower animals, and animals may drink in laboratories for reasons inapplicable to man. However, when excessive quantities of alcohol are self-administered for whatever reason, it is well to know how intoxication, tolerance, withdrawal and dependence in humans and lower animals compare.

Tolerance and withdrawal

Tolerance is the reduction of a drug's effects with repeated administration; withdrawal is induction of the opposite effects of a drug when the drug is no longer taken. Tolerance is not a sign of physical dependence, but withdrawal symptoms are because they are removable by further drug administration. With alcohol, which is a depressant, tolerance is evident through a number of behavioural changes, including sleep times with prolonged administration; withdrawal is exemplified by symptoms that include hyperactivity and, eventually, convulsions.

Tolerance is exhibited either when a constant drug dose produces a diminished effect or when a constant effect is produced with an increased dose. These alternatives are shown in Figure 6.1, where A and B are hypothetical dose–response (effect) curves from early and late drug administrations, respectively. Points X and Y show a diminished response at the same dose level, P, and points X

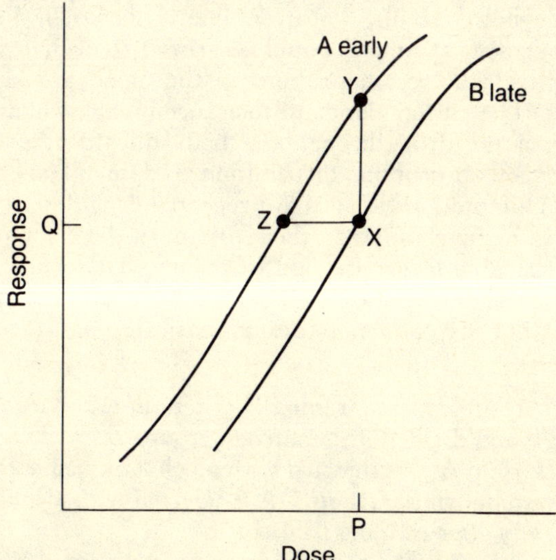

Figure 6.1 Hypothetical dose-response curves illustrating changes from early use (A) to late use (B). At the same dose level, P, the drop from X to Y illustrates tolerance by diminished responses to a constant dose, and at the same response level, Q, the shift from X to Z illustrates tolerance by increased dose for a constant effect.

and Z show the corresponding drug doses that achieve the same effect, Q, from the early to the late drug administration.

Investigation of tolerance with constant doses cannot employ voluntary drug administration because dose level then is beyond experimental control. Consequently pharmacologists typically demonstrate tolerance acquisition in animals by injecting predetermined doses and recording effects as a function of successive injections. The effects may take a variety of forms of physiological and behavioural measures (Kalant, LeBlanc and Gibbins, 1971). With rats, typical signs of acute alcohol intoxication used as laboratory measures for the study of tolerance development are loss of righting reflex, diminished ability to climb a pole, increased difficulty in clinging to a tilted wire frame, and reduced ability to keep from falling off a moving belt. Tolerance, however, does not develop at the same rate for all behaviours, and it possibly occurs faster the more important the behaviour is for the animal, such as procurement of food.

Three kinds of tolerance to alcohol have been identified in animals, one acute and two chronic. The acute kind, called tachyphylaxis or the Mellanby effect, is shown by a greater impairment at equal blood alcohol levels when the level is rising than when it is falling after a single alcohol experience. The course of blood alcohol concentration over time is shown by Figure 6.2, where impairment at time P would be greater than that at time Q, even though alcohol concentrations in the blood are identical at these times. The chronic kinds of tolerance involve changes in the disposition of the drug through the body due to adjustments in mechanisms of absorption, distribution and metabolism, and in behavioural adaptations to constant repeated drug doses over time.

Motorists often claim that their driving is less impaired than their blood alcohol levels are said to predict. The claim is usually exaggerated, but it does have a basis in laboratory animal behaviour. The experimental design basically has three phases (Chen, 1968).

Phase 1. All subjects are trained to criterion, tested for alcohol sensitivity and divided into matched groups.
Phase 2. Group A-T is injected with alcohol immediately before training trials; Group T-A is injected with alcohol immediately after training trials.
Phase 3. Test trials are run with both groups after alcohol administration.

With more or less complex variants of this basic paradigm workers constantly find more rapid tolerance development in animals given alcohol immediately before training trials (A-T) than in animals

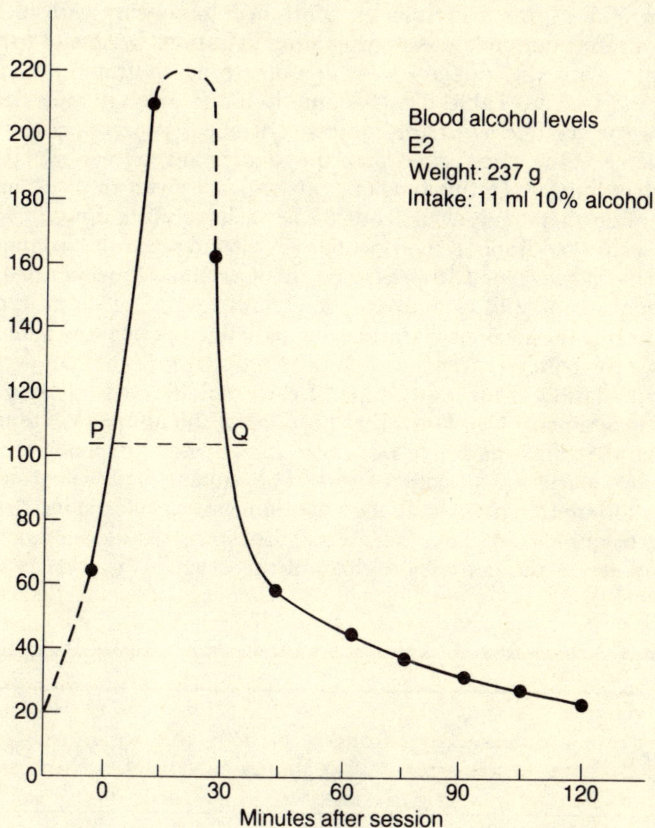

Figure 6.2 Blood alcohol levels in milligrams of alcohol per hundred millilitres of blood in a hungry 237 g white rat that had ingested 11 ml of 10 per cent alcohol (w/v) in a 30-minute session in which drinking was reinforced with food. Points joined by solid lines were measured by gas chromatography from tail blood samples; dotted lines are estimates. Acute tolerance (tachyphylaxis) would be demonstrated by different effects at the rising (P) and falling (Q) points of equal blood-alcohol levels.

given equal amounts immediately after training (T-A). This phenomenon is also known as behaviourally augmented tolerance, but conditioning contributions to ethanol tolerance and dependence may be more than just augmentation (Hinson and Siegel, 1980).

Most studies of the development of tolerance in animals use the *constant dosage* (or experimenter-controlled) procedure with intubation or injection methods of drug administration. Results of various studies with this method with non-opiate psychotropic drugs are summarized in Table 6.2 (Kalant, LeBlanc and Gibbins, 1971). However, *constant effect* (or subject-controlled) procedures are also possible. One such case pits the depressant effects of alcohol against hunger. In this type of study rats are given short rations to maintain their body weights at 85 per cent of normal, and trained to secure additional food pellets by licking a tube containing saccharin-sweetened 10 per cent alcohol solution. Once trained, the animals drink and eat rapidly for various lengths of times in daily experimental sessions with shorter or longer periods of inactivity between bouts of drinking and eating. Figure 6.3 shows the accumulation of inactivity time at early, middle and late stages of the experiment. The horizontal portions of the figure are the times when the animal is actively engaged in eating and drinking and the vertical parts are periods of rest. The animal plainly drinks and eats for shorter periods at the beginnings of early sessions than at the beginnings of later sessions, suggesting the development of tolerance to the depressant effect of the drug.

Table 6.2 Development of tolerance to psychotropic drugs in various species

Drug	*Species*	*Time of Measurement*	*Extent of tolerance*	*Reference*
			%	
Barbiturate	Dog	6weeks	43–57	215
	Dog	2days	50	97
	Dog	4–7days	60–100	135
	Dog	5–7days	43-64	138
	Guinea pig	4weeks	60	32
	Mouse	5–6days	50	156
	Mouse	5days	23–27	320
	Mouse	4–14days	150–327*	115
	Rabbit	2–3days	50	103
	Rabbit	4–7days	27–35	254
	Rabbit	3–5days	36–47	138
	Rabbit	3days	67	124
	Rat	2–3days	100	288
	Rat	2days	43–50	277
	Rat	5days	15–57	138
	Rat	1–3days	11–40	9

Table 6.2 continued

Drug	*Species*	*Time of Measurement*	*Extent of tolerance*	*Reference*
	Rat	1day	26–55	10
	Rat	4days	21–22	383
	Rat	35days	46	383
	Man	80days	'Virtually complete'	165
	Man	30–70days	100	211
Major tranquilizers	Dog	15days	53	387
	Mouse	2–6days	60	218
	Mouse	1–4weeks	60–81	368
	Rat	12weeks	62	25
	Rat	14days	'Almost complete'	159
	Rat	15days	80	256
Minor tranquilizers	Cat	21–50days	60	401
	Rat	35days	44	301
	Rat	5days	—	155
	Rat	14days	42-71	128
	Rat	9–35days	58–67	255
	Mouse	6–11days	50–75	38, 39
Stimulants	Rat	1–21days	19	333
	Rat	13–30days	75	331
	Man	14days	Complete	316
Hallucinogens	Rabbit	4days	54	125
	Rat	7–8days	100	113
	Rat	7days	83	352
	Rat	2–7days	70–100	6
	Man	7–21days	0–87	162
	Man	6–13days	Complete	166
	Man	14days	33–100	315, 316

* In this experiment, tolerance was determined as the change in drug dose or concentration of drug in the serum which provided 50 per cent protection against a standard electroconvulsive stimulus. In all other cases cited, tolerance is given as the reduction in effect produced by a given dose, so that the maximum tolerance cannot exceed 100 per cent. Reference numbers refer to the original publication. From Kalant, H., LeBlanc, A.E., and Gibbins, R.J. (1971) 'Tolerance to, and dependence on, some non-opiate psychotropic drugs', *Pharmacological Reviews*, 23, 135–91. Copyright 1971, The Williams & Wilkins Co., Baltimore. Reprinted by permission.

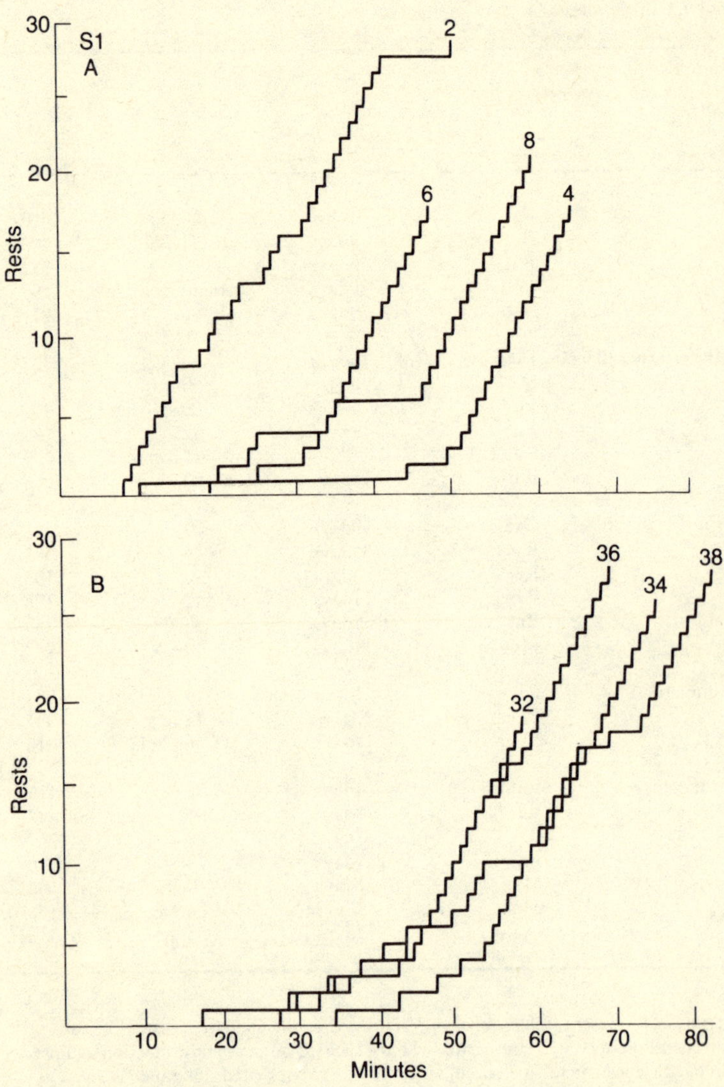

Figure 6.3 Result of an experiment in which hungry rats licked a metered alcohol solution for food reward. Drinking and eating are sustained at the beginning of sessions (horizontal lines) after which periods of sleep or inactivity accumulate (vertical lines). In early sessions eating and drinking are interrupted by sleep earlier than in later sessions, demonstrating tolerance to inhibitory effects of self-administered alcohol with experience. (Data collected by Arnold N. Mayers.)

A procedure such as this is unsatisfactory for the procurement of traditional pharmacological dose–response data because daily dose levels are uncontrollable, but it is a closer imitation of the human method of drinking than is the pharmacological procedure inasmuch as it is the drinker, not an external agent, who determines how much is drunk. A compromise procedure is that of incorporating alcohol into the animal's diet to meet a predetermined amount of its daily calorific requirement. Freund (1969, 1980) used this method to study the alcohol withdrawal syndrome in mice, and it is also useful for investigating the foetal alcohol syndrome (Bond and DiGiusto, 1977). With it, it is possible to predetermine either the percentage of ethanol-derived calories an animal consumes per day, or the total number of grams per kilograms of alcohol it ingests.

Alcohol withdrawal syndromes in mouse, monkey, man, dog and rat

The method of incorporating alcohol into an animal's diet is hardly equivalent to permitting the animal to take the beverage or leave it. Nevertheless it is a benign way of delivering alcohol to experimental subjects in controlled amounts. It does not involve the stress of intubation or injection and can provide standard doses for comparing pharmacological and behavioural effects of alcohol on animals and man. With this method, tremors, stereotyped back-up behaviours and convulsions have been found after alcohol-withdrawal from mice (Freund, 1969), and rapid respiration, hyperreflexia, mild tremulousness, irritability, spastic rigidity, sweaty palms and feet, and convulsions have been shown to comprise the alcohol withdrawal syndrome in young chimpanzees (Pieper, Skeen, McClure and Bourne, 1972). These compare with the autonomic (perspiration, fever, elevated blood pressure, nausea, vomiting and diarrhea) and central (tremor, hyperreflexia, agitation, insomnia, hallucinations, convulsions and delirium) nervous system manifestations of alcohol withdrawal in man (Isbell *et al.*, 1955), and with the successive stages of withdrawal from alcohol that human alcoholics display.

Stage 1. Tremors, excessively rapid heart-beat, hypertension, heavy sweating, loss of appetite, and insomnia.
Stage 2. Hallucinations, auditory, visual, and tactile, singly or in combination.
Stage 3. Delusions, disorientation, delirium, sometimes intermittent in nature and usually followed by amnesia.
Stage 4. Seizure activity (Ray, 1978).

In rhesus monkeys given up to 8 grams per kilogram of 25 per cent

Table 6.3 Ethanol withdrawal reactions in rhesus monkeys

Stages and Reactions	*BEC*	*Observations of Reactions in Monkeys*							
		1		*3*	*4*	*5*	*6*		
		A	*B*				*A*	*B*	*C*
	mg/100 ml								
Tremulous	150–200								
Generalized tremors		X	X	X	X	X	X	X	X
Muscle fasciculations		X	X	X	X	X	X	X	X
Elicited hyperreflexia		X	X	X	X	X	X	X	X
Spastic	50–150								
Spasticity		X	X	X	X	X	X	X	X
Rigidity		X	X	X	X	X	X	X	X
Spontaneous hyperreflexia		X	X	X	X	X	X	X	X
Behavioural changes		X	X	X	X	X	X	X	X
Apparent fright		X	X	X	X	X	X	X	X
Salivation				X	X	X			
Mydriasis				X	X	X	X	X	X
Retching/vomiting			X		X				
Spastic or convulsive 'poses'		X	X						
Convulsive	20– 85			X (6) Died	X (2)	X (5)	X (2)		X (2)
Clonicotonic convulsions									

Observation of reactions in individual animals is indicated by X. Blood-ethanol concentration corresponding to each stage is given as general ranges. The number of convulsions occurring in each monkey is noted in parentheses. A, B and C refer to separate experiments. From Ellis, F.W., and Pick, J.R. (1970) 'Experimentally-induced ethanol dependence in rhesus monkeys', *Journal of Pharmacology and Experimental Therapeutics*, 175, 83–93. Copyright 1970, The Williams & Wilkins Co., Baltimore. Reprinted by permission.

(weight by volume) alcohol by gastric intubation, tremors, spasticity and convulsions in that order occurred after alcohol withdrawal (Ellis and Pick, 1970). Table 6.3 shows the full range of symptoms exhibited by the six experimental monkeys, and the blood alcohol levels associated with them. Ellis and Pick (1970) characterize the stages of intoxication and withdrawal signs shown by their animals over 10 to 18 days of alcohol treatment and 2 to 3 days of withdrawal as:

> *Chronic intoxication*. Sedation, lethargy, ataxia, uncoordination and sleep at first, followed by fine tremors in the fingers when blood alcohol measures declined.
>
> *Withdrawal stages*. *Tremulous stage, including hyperreflexia*; *spastic stage*, characterized by spontaneous body twitches and jerks; *convulsive stage*, with sustained tonic and clonic seizures.

An important conclusion drawn from this study is that withdrawal signs are associated with the decline in blood alcohol concentration from a critical level rather than with the disappearance of alcohol from the body, or the time since alcohol was last ingested.

With dogs, too, tremors, sleeplessness, convulsions and apparent hallucinations occur after withdrawal from alcohol delivered by gastric intubation (Essig and Lam, 1968). With the same delivery method, Majchrowicz (1975) found the reactions of experimental rats to increasing (intoxicating) and decreasing (withdrawal) levels of alcohol in the blood that are summarized in Figure 6.4.

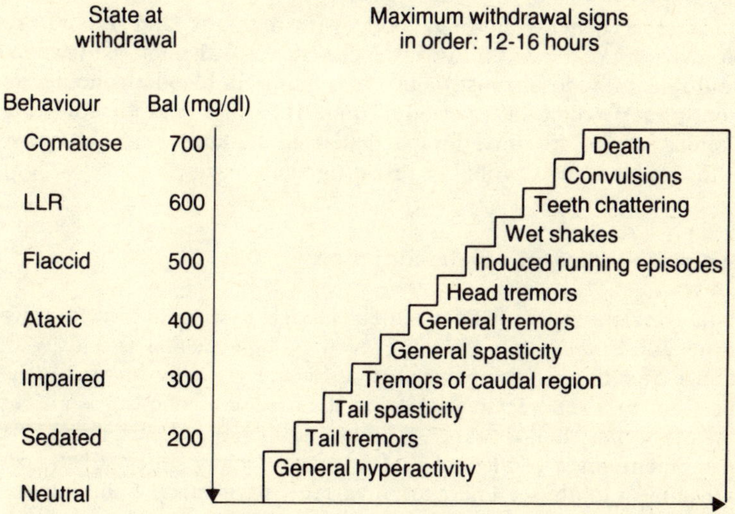

Figure 6.4 Ethanol dependence in rats as indicated by withdrawal signs over a sixteen-hour period after the last of a series of intragastic ethanol intubations (20% w/v) spread over a four-day period. For details see Majchrowicz, E. (1975) 'Induction of physical dependence upon ethanol and the associated behavioral changes in rats', *Psychopharmacologia* (Berlin), 43, 245–54.

Diet manipulation, gastric intubation and intravenous injection are all methods for administering alcohol in controlled and sufficient quantities to induce in animals alcohol intoxication and withdrawal syndromes similar to those observed in human drinkers. However the intermittent feeding procedure that generates schedule-induced water polydipsia in rats has the advantage of inducing high levels of alcohol intake by self-administration. With this method, Falk, Samson and Winger (1972) utilized multiple daily experimental sessions 1 hour long and 3 hours apart in which food pellets were delivered to hungry rats at intervals 2 minutes apart. With only a 5 per cent alcohol solution to drink in experimental sessions the animals consumed enough to maintain voluntary alcohol intakes around 13 grams of alcohol per kilogram of body weight. When alcohol was withdrawn, evidence of dependence appeared in the forms of audiogenic seizures to the sound of jangling keys, spontaneous convulsions, and death. In home cages with single daily food rations and 5 per cent alcohol available for drinking, similar animals drank substantially at self-selected mealtimes, but no signs of physical dependence on alcohol developed (Samson and Falk, 1974).

It is a general finding, both with humans and laboratory animals, that for withdrawal-dependence on alcohol to develop, multiple daily doses must be taken to maintain blood alcohol levels continuously high for a period of time. How high and for how long cannot yet be specified for particular individuals, man or beast, although they will probably rest on genetic bases.

Animal models in opiate addiction

It is convenient to consider animal models in alcohol and in other drug abuse separately, because even though alcohol is a drug of abuse and the requirements of models are virtually identical, the need for oral self-administration in the case of alcoholism is absent in the other major case – opiate addiction. Apart from the prominent cases of alcohol and narcotics, many other substances are subject to abuse. These are invariably psychotropic and may be classified in various ways depending on chemical, physiological, psychological or social considerations. A useful compromise is that of Ray (1978) who lists alcohol, nicotine and caffeine as non-drugs like other over-the-counter and prescription preparations, and divides the remaining psychotropic substances into three main groups by either their *psychological effects* – narcotics and phantasticants (hallucinogens, marijuana and hashish) – or by their *therapeutic uses* (tranquilizers, stimulants and depressants).

A simpler classification embraces four categories:

1 stimulants and depressants;
2 major and minor tranquilizers;
3 hallucinogens; and
4 opiates.

Caffeine, amphetamine, alcohol and barbiturates are members of the first category; anti-psychotic and anti-anxiety drugs fall in the second; LSD, mescaline and atropine belong in the third group; and morphine and heroin are members of the fourth. Substances from all these categories are ingested by free-ranging animals and many have been subjects of tests with laboratory animals. In the case of therapeutic uses of drugs in psychiatry, many of which were accidental discoveries, animal models are employed for tracing a drug's mode of action and also for assessing a new drug's liability to abuse. This latter entails measuring the power of a drug as a reinforcer and testing its capability to addict and readdict.

Drugs as reinforcers

Reinforcement is a technical term in psychology originally used by Pavlov (1927) to describe brain processes thought to occur during conditioned and unconditioned stimulus associations. Drugs-as-reinforcers in this sense refers to them as unconditioned stimuli that first elicit unconditioned responses and then reinforce these responses when they are conditioned to previously neutral stimuli. Pavlov recounts such a case where the response pattern of profuse salivation, nausea, vomiting and sleep (UCR), elicited in a dog by hypodermic injection of morphine (UCS), was eventually elicited as a conditioned response pattern (CR) during preparation of the injection syringe (CS). Actual morphine injection is essential to maintain the conditioned behaviour, and it is in this sense that morphine is called a reinforcer.

The term is used in the Pavlovian sense below in connection with an account of opiate tolerance and withdrawal, but in the present section the operant use of the term is employed. In this case a reinforcer has the characteristic of a reward insofar as it is a stimulus consequence of a procuring behaviour. If the behaviour is learned, or occurs more rapidly when followed by the stimulus than otherwise, the stimulus is defined as a positive reinforcer. In this sense, if an animal learns to self-administer a drug, the drug is a positive reinforcer. When a stimulus decreases response strength, the stimulus is a punisher; and if a new response is learned to escape or avoid it, the stimulus is called a negative reinforcer.

A pre-requisite for the study of drugs as operant reinforcers is the availability of an automatic system for intravenous self-administration. Figure 6.5 shows such a system, based on one devised for the rat (Weeks, 1962), used with the monkey by Schuster and Johanson (1974). With such equipment, cocaine, amphetamine, methamphetamine, nicotine, short-, medium- and long-acting barbiturates and morphine have all been voluntarily self-administered by rats and monkeys (Schuster and Johanson, 1974).

Because of the aversive consequences of morphine withdrawal it is questionable whether positive or negative reinforcement is responsible for sustained self-administration of morphine. An early

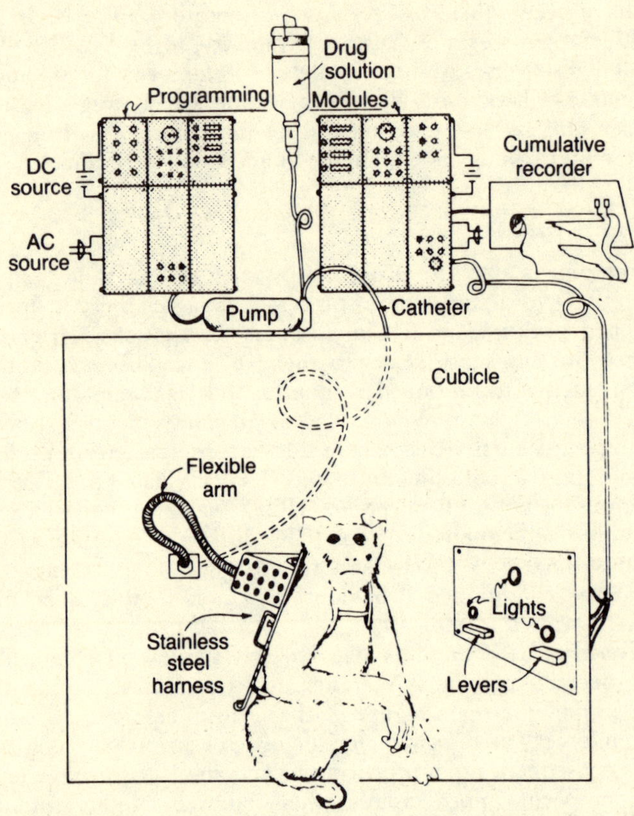

Figure 6.5 Apparatus devised for intravenous self-administration of drugs by laboratory monkeys. (From Schuster, C.R., and Johanson, C.E. (1974) 'The use of animal models for the study of drug abuse', in R.J. Gibbins [editor-in-chief], *Research Advances in Alcohol and Drug Problems*, Vol. 1, New York, Wiley, 1974, Copyright 1974 by Werbel & Peck, Publisher. Reprinted by permission.)

study with rats (Weeks, 1962) employed prior forced injections to make the animals morphine dependent, and then showed that they learned to barpress for more morphine injections. This establishes morphine as a negative but not necessarily as a positive reinforcer. Later work with monkeys, however, has employed morphine doses too low to make self-administering animals physically dependent, and has demonstrated positive reinforcement by the drug.

Some representative drugs self-administered by laboratory animals are narcotic opiates (codeine, heroin, methadone, morphine), psychomotor stimulants (cocaine, amphetamine, but not caffeine), central nervous system depressants (chlordiazepoxide, alcohol, pentobarbital) and the anaesthetic and one-time street drug, phencyclidine. Those drugs not self-administered are narcotic antagonists that generate withdrawal symptoms (naloxone), anti-depressants, major tranquilizers and hallucinogens (Schuster, Renault and Blaine, 1979). 'It is our contention', Schuster *et al.* conclude, 'that these animal data strongly suggest that from a biological viewpoint drug-seeking behavior is normal.' They continue:

> This would prompt us to paraphrase the later John B. Watson and say – give me twelve healthy babies and control over their environment and I can make any one of them a speed freak, a heroin addict, a barbiturate fiend, etc. (Schuster *et al.*, 1979, p. 5)

Control over the environment, in the form of reinforcement dispensation, is not only a pre-requisite for the initiation of drug use but also for the establishment or prevention of readdiction. In the case of initiation the reinforcement is operant; in the case of readdiction, Pavlovian reinforcement is involved.

Opiate readdiction: conditioned tolerance and withdrawal

Behavioural, or behaviourally augmented tolerance, described above for alcohol, has a parallel in conditioned tolerance with morphine (Siegel, S., 1979). Apart from its euphoriant effects, morphine is an analgesic, and this property is studied in laboratory rats with a so-called 'hot plate' technique. With this method, a morphine-dosed subject is placed on an electrically heated surface and the point is noted at which it lifts a paw as the current is increased. With repeated exposure to the procedure, paw lifting occurs sooner and sooner, showing decreasing analgesic potency of the drug, or tolerance. Conditioned tolerance to morphine with this method has been demonstrated with three experimental designs (Siegel, S., 1979).

1 *Environmental specificity design.* Two groups of rats are given morphine injections in environments E_1 (group S) and E_2 (group D). The hot-plate tolerance tests are conducted in E_1, such that group S is injected and tested in the same environment while group D is injected and tested in different environments. More tolerance is shown by group S than by group D, indicating environmental control of tolerance over and above physiological effects.

2 *Signalled* versus *unsignalled injection design.* Two groups of rats

experience an audiovisual signal and morphine injections. For group A-V the signal and injections are paired; for group N they are not. Both groups are tested on the hot plate for tolerance with the signal and injection paired, so that injection and test conditions are the same for group A-V and different for group N. Group A-V exhibits conditioned tolerance; group N shows none.

3 *Discriminative control design*. All subjects receive injections of morphine in environment E_M on some days and of physiologically neutral saline in environment E_S on other days. Hot plate tests for tolerance are conducted in E_M for half the rats (group E_M) and in E_S for the others (group E_S). Group E_M, the same-tested rats, but not group E_S, developed morphine tolerance.

With all three experimental designs, tolerance to the analgesic property of morphine developed only when injection and testing conditions were the same. When tests for tolerance were made under novel environmental conditions, tolerance for the drug was not displayed. This finding has important implications for the explanation of drug addiction and relapse. The argument involves three steps.

1 An addicting drug has two physiological effects, one of a direct reaction, the other an opposing process, a kind of rebound or neutralizing effect.

2 The direct effect is conditionable as described above, but the magnitude of the unconditioned response is fixed. The neutralizing effect is also conditionable and grows with repeated administration of the drug. The net effect is a diminished drug response, which is conditioned tolerance.

3 In the absence of the drug after the rebound response is conditioned (i.e. a Pavlovian extinction trial for the direct effect), only the rebound response occurs in the drug administration environment. This is conditioned withdrawal, which is responsible for the readdiction, or relapse, of recovered morphine addicts when they return to the environment where they first became addicted.

Figure 6.6 (Siegel, S., 1979) is a graphic depiction of the process.

Conditioned morphine withdrawal has been demonstrated directly with monkeys (Goldberg and Schuster, 1967) and with human volunteers (O'Brien *et al.*, 1977). The monkeys were made morphine dependent by intravenous infusion, whereafter a withdrawal reaction was evoked with the morphine antagonist, nalorphine. A buzzer accompanied nalorphine administration, and

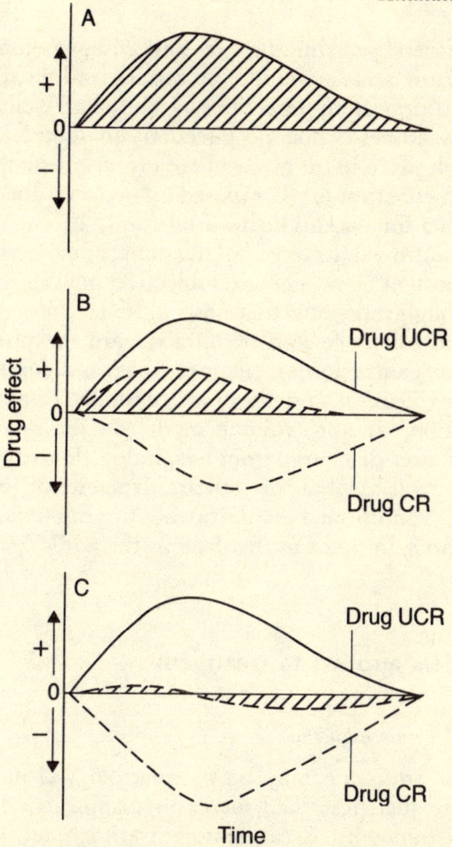

Figure 6.6 Time course of the net drug effect (hatched area) resulting from the interaction between the pharmacological Unconditioned Response and the drug-compensatory Conditioned Response during successive stages in tolerance development.

A Time-effect curve of a drug on initial administration.

B Smaller net effect after several administrations due to compensatory effect of the drug CR.

C Paradoxical effect after many administrations leading to maximum compensatory drug CR. This effect occurs with long-term opiate addicts and animals given many morphine injections.

From Siegel, S. (1979) 'The role of conditioning in drug tolerance and addiction', in J.D. Keehn (ed.), *Psychopathology in Animals: Research and Clinical Implications*, New York, Academic Press, pp. 143–68, reprinted by permission.

on test the buzzer alone evoked the withdrawal response. The humans were addicts clinically maintained with daily doses of the morphine substitute, methadone. For them, the withdrawal syndrome, consisting of running eyes and nose, yawning, decreased skin temperature and increased breathing and heart rates, was evoked by intramusculur injections of the short-acting morphine antagonist, naloxone. Naloxone injections (or blind control saline injections) took place in an atmosphere of soft music that gave way to a conditioned stimulus composed of a tone and peppermint odour. On tests for conditioned withdrawal, all subjects reported subjective withdrawal feelings to the tone and odour compound alone, and most of them showed objective physiological signs as well, even though naloxone was not injected.

Conditioned tolerance and withdrawal are the probable mechanisms behind readdiction, or relapse, which is a common problem following detoxification and hospital treatment of alcoholism and opiate addiction. If the animal model is appropriate then it indicates that after discharge from hospital or detoxification centre, the treatment of alcohol or opiate dependence must include extinguishing conditioned withdrawal by administration of a neutral substance in place of the drug in the addict's normal social environment.

Animal models applied to treatment

Methadone and opiate addiction

Treatment for addiction may be pharmacological or behavioural. The procedure just described for extinguishing conditioned withdrawal is a behavioural treatment, but with opiate addicts most interest now is on pharmacological treatment by methadone maintenance. This treatment is not expected to cure addiction but to substitute the cheaper and less potent methadone for heroin or morphine, with the aim of making the addict less dangerous to society. The aim was the same when heroin was introduced in 1898 as a substitute for morphine.

Methadone is a prosthetic device in the treatment of opiate addiction. It permits the addict to function in society, just as a wheelchair helps the lame or spectacles help the nearsighted. A cure would make morphine unattractive or aversive, but a recent review of methadone treatment programmes for narcotic addicts concludes that it is not so much pharmacologically-induced euphoria that maintains drug-taking behaviour as social and psychological factors.

> The bulk of the evidence, we believe, indicates that addiction is a disorder profoundly reinforced in its *initiation* by

> social–psychological as well as pharmacological factors. 'Friends' group together and employ each other to be different, 'cool' and daring. Addiction is then sustained by social and psychological factors in which the reinforcing value of narcotic drug plays an important, but not an essential role. Whether tranquilizing and antianxiety drug effects are contributory motives in sustaining addiction is as yet unclear; but we do know that euphorigenic effects do decrease and in some instances disappear entirely as the dependence grows older. We do not think that physical dependence itself is sufficient to generate the profoundly self-destructive behaviors characteristic of the modal pattern of addiction with which the public is most concerned. (Freedman and Senay, 1973, p. 155)

The need for an animal model of psychopathology in a social context has been recognized (Ellison, 1979) but has not been applied to treatment of opiate addiction or alcohol abuse. Ellison has, however, shown that social drinking by rats under certain conditions in a large laboratory enclosure resembles the 'cocktail hour' and 'nightcap' patterns of human drinking.

Pharmacological treatments of alcoholism

With alcoholism, there are two classes of treatment with bases in animal studies, a pharmacological class and a behavioural class, both with ambitions to cure. Each class has two different theoretical foundations.

The pharmacological treatments of alcoholism capitalize, respectively, on the biochemistry of alcohol metabolism and on drug-induced relaxation. The first aims to make alcohol consumption aversive; the second to reduce tensions supposedly responsible for drinking. Disulfiram is a member of the first category. It prevents the elimination of metabolites of alcohol from the body so that violent sickness follows alcohol consumption during disulfiram treatment. Mottin (1973) describes some dangers with this treatment, as well as regimens that have had some success. Disulfiram succeeds with alcoholics already determined to discontinue drinking, otherwise they stop taking the drug. It is not a therapeutic drug but is a self-administered threat of punishment for drinking alcohol. As such, it is a psychological method of aversion treatment (see below) based on a biochemical effect specific to alcohol, and can be considered a prosthetic device as much as a therapeutic agent.

Ataractics are non-specific drugs for alcohol drinking. They are anxiety-reducers and can diminish alcohol consumption only so far as alcohol is consumed for anxiety-relief. As they too can become abused they stand in relation to alcohol as methadone stands to opiate abuse.

Behavioural treatments of alcoholism

Three forms of behavioural treatment are used with alcoholics. Each evolved from laboratory studies of learning in animals, the first two from the tradition of Pavlov, the third from that of Skinner. They are, respectively, aversive respondent conditioning therapy, systematic desensitization, and the positive reinforcement of alternative operant behaviour to drinking. The first of these makes alcohol aversive; the second is the behavioural equivalent to ataractic drug therapy; and the third seeks to instil controlled drinking.

The rationale for aversion therapy is that the sight, smell, taste and other paraphernalia associated with alcohol consumption are conditioned stimuli (CS) for a conditioned pleasurable response (CR) evoked by alcohol consumption. The objective of this treatment is to condition aversion instead of pleasure to the alcohol-drinking situation. Originally this was done with an emetic administered prior to stimulation with alcohol. If the alcohol stimuli are presented just as the emetic is effective, then nausea, not pleasure, is evoked by the alcohol situation. Because of uncertain latencies of emetic actions, and improper use of conditioning procedures, the timing of stimulus presentations in clinical situations has often been erratic, and therapeutic results unreliable. A move to replace emetine by electric shock aversion, which can be timed precisely, has not achieved popularity because, unlike vomiting, shock aversion is not a natural consequence of alcohol ingestion.

Systematic desensitization, or reciprocal inhibition, is a classic form of behaviour therapy devised as a treatment for experimental neurosis in cats (Wolpe, 1958), although an earlier laboratory analogue was employed with a fearful child (Jones, 1924). The technique involves establishing a hierarchy of circumstances that evoke a particular behaviour. As relaxation is conditioned to the least anxiety-raising situation, so conditioned relaxation to the next step in the hierarchy is attempted, and so on, step-by-step, until not anxiety but relaxation occurs in all situations. As with ataractic (anxiety-reducing) drug therapy, the approach is relevant only in those cases when anxiety-reduction is the reason for excessive drinking.

Several forms of operant therapy, or behaviour modification, are employed with alcoholics, from limited behaviour contracting with individuals or families to total involvement with therapeutic communities (Keehn *et al.*, 1973). The principal derivations from laboratory-based operant theory are the contingency control of behaviour on the one hand and the stimulus control of behaviour on the other. These involve diagnosis, in the sense of discovering the circumstances under which a person drinks (stimulus control), and the payoffs that excessive drinking brings (contingency control) in situations that, to the outside observer, seem aversive. When these are found, contracts are drawn up between a therapist and a client, or between husband and wife (Miller, 1972) such that the client loses money deposited with the therapist if he breaks his side of the contract, or gains some if he refrains from drinking except under mutually agreed conditions.

7 True and model psychosis

Psychotic symptomatology

In humans

Psychosis is a technical term of medical psychology that describes a condition variously known as lunacy, madness or insanity. Insanity is not a psychological term but a legal one. It is not exactly equivalent to psychotic illness, for insanity legally refers to ignorance of the effects of one's action. Psychotic individuals may know what they are doing but be unable to judge the significance of their behaviour, a presumably normal characteristic of lower animals. Madness is also non-technical, psychologically speaking, and is generally applied to senseless behaviour often accompanied by violence. The insane and the mad are without the full complement of human faculties and in this sense resemble animals, but mad dogs are not models of human psychosis. Lunacy is a legacy of a discarded theory that madness depends on the phase of the moon.

There are many signs of psychosis in humans, although not every one occurs in every patient. Figure 7.1 shows the frequency of different kinds of disturbance found among 500 consecutive admissions to an American hospital for mental disorders. The list includes emotional apathy, thought disturbance, anti-social behaviour and various other signs that the patients are not in touch with the reality of the world around them. Psychotic behaviour is peculiar and often annoying, and patients can be a danger to others and to themselves. Most patients are emotionally upset in one way or another, and many report experiences that to others are unreal.

According to their symptom patterns, human psychotics are usually divided according to the diagnostic categories listed in Chapter 1. The major categories are the schizophrenias and affective disorders, each with various subdivisions; the minor

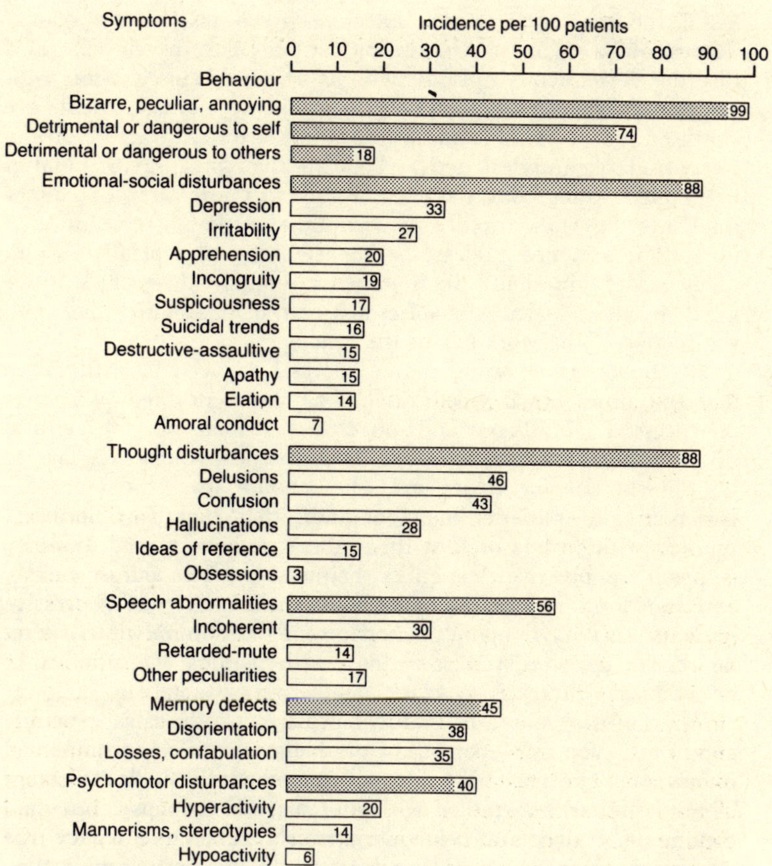

Figure 7.1 Symptoms shown by 500 consecutive admissions to an American psychiatric institution. (From Coleman, James C. (1950) *Abnormal Psychology and Everyday Life*, Copyright 1950 by Scott, Foresman & Company, Reprinted by permission.)

categories are paranoid disorders and involutional psychosis. Paranoid disorders typically involve delusions as their major defining feature, mainly of persecution and grandeur. The other minor category, involutional psychosis, has as its main characteristic depression. The diagnosis of involutional psychosis refers to depressive psychosis which first appears during the involutional period, and the feeling of depression is frequently associated with self-recrimination. Such patients typically brood over real or imaginary failures in their lives, or unpardonable sins they have committed.

Of the major psychoses, affective psychoses include mania, depression and manic-depression or circular psychosis. This disorder is frequently episodic with short manic episodes interrupting long periods of depression. During acute mania, patients are confused as to who and where they are, and are typically uncontrollable and destructive. They are the typical raving lunatics of popular fancy, but patients of this sort are rare. In calmer moments they talk rapidly in a continuous flow of disconnected ideas that are meaningless to the listener, and typically exhibit delusions of grandeur. Such patients imagine themselves to be great inventors, scientists, statesmen or religious figures that hold the secrets of the workings of the universe.

At the other extreme, manic-depressives become withdrawn, dejected, immobile and bedridden. They are bedevilled by feelings of excessive sinfulness and often feel responsible for natural disasters like earthquakes, droughts and floods, which they take to be punishment for their sins. Hypochondriacal tendencies are common, and patients may complain that their intestines are infested with worms or that their stomachs have turned to stone. Typically, depressives feel guilty, helpless, hopeless and unworthy, and suicide attempts are common. About a quarter of depressive patients also have manic periods, sometimes immediately after depressive episodes and sometimes with periods of normality in between.

The natural course of manic-depressive psychosis is generally good for younger persons, although recurring attacks of mania or depression are more common than single episodes. Thanks in part to studies with animals, affective psychoses are coming under drug and behavioural control. The same may be true of schizophrenia which, in the past, had a poorer prognosis than affective psychosis.

Schizophrenia, once called dementia praecox to call attention to its characteristic of precocious mental deterioration, is the most common psychological disorder found in mental hospitals. Until recently, typical schizophrenics were doomed to progressive mental deterioration over the years, but modern drug treatments appear to have stopped this trend. Nevertheless, the condition remains largely as it was described by Osmond and Smythies in 1952.

> A situation in which a person of any age, but usually a young adult, in response to stress or with little evidence of it, becomes slowly and insidiously, or with overwhelming speed and accompanied by acute confusion, subjected to disturbances of association, changes in affect, thought disorder, hallucination and delusion and catatonic symptoms

> to such an extent that life outside a mental hospital becomes impossible. The sick person, on the other hand, may never need to visit a doctor but may simply appear odd and eccentric. The illness may terminate quickly either with or without medical aid, or may be completely resistant to any form of treatment and continue for years without any pathognomic physical changes being demonstrable.

The passage succinctly describes the manifestations of schizophrenia in disturbances of perception, mood and movement, but it also dramatically highlights the unpredictability of onset and remission. Onset is as unpredictable now as it was when the passage was written, but it is with remission that conditions have changed, largely on account of pharmacological treatments and studies with animals.

Emotional flatness and apathy are characteristics of schizophrenia. The term means split mind, where splitting refers not to split, or multiple, personality, as is commonly supposed, but to separation of emotional and intellectual psychological functions. Intellect as well as emotion is abnormal. McKellar (1957) likens schizophrenic thinking to the knight's move in chess – peculiarly crooked. For example, the question 'Where is Egypt?' may be answered, 'Between Babylon and the Congo', and the interpretation of 'A bird in the hand is worth two in the bush' given as 'That could be your bank account as opposed to your income tax.' Both replies have a certain logic, yet neither one is quite normal. Another characteristic of schizophrenic thinking McKellar calls loss of the 'as if'. The phenomenon has been noticed in drug-induced psychosis. One medical experimenter, after a dose of an impure adrenaline solution, hallucinated patterns that assumed the shapes of fishes. The report of his experience reads, in part, 'I felt I was at the bottom of the sea or in an aquarium among a shoal of brilliant fishes. At one moment I concluded I *was* a sea anemone in the pool' (Hoffer, Osmond and Smythies, 1954).

Schizophrenics also suffer from delusions. Body image delusions include belief that the body is extraordinarily large or small, or that it is rotting away, or that it has turned to wood or stone. Patients may have delusions of grandeur, as in the case of affective psychoses, or delusions of reference wherein they see sinister references to themselves in the everyday acts of other people, as in paranoia. Sutherland (1976) describes a typical case of paranoia in a fellow-patient.

> One night, when I was telephoning, he hung around interrupting every word. He heard me saying: 'I'll see you tomorrow,' and was convinced I was in a plot against him. 'I

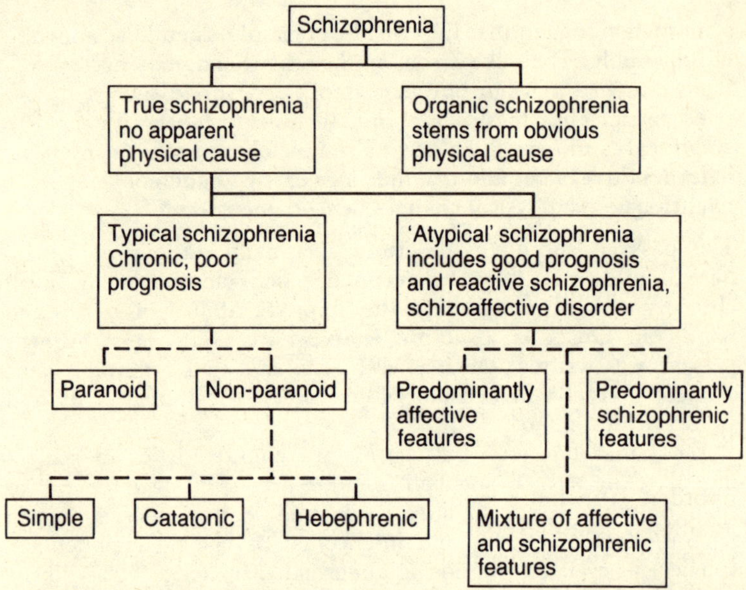

Figure 7.2 A schematic representation of the subgroups of schizophrenia currently accepted by most psychiatrists. Solid lines connect subgroups for which there is established evidence; broken lines connect subgroups for which there is unverified or incomplete evidence. (From Tsuang, M.T. (1982) 'Schizophrenic syndromes: The search for subgroups in schizophrenia with brain dysfunctions', in F.A. Henn and H.A. Nasrallah, *Schizophrenia as a Brain Disease*, New York, Oxford University Press, pp. 14–25, Reprinted by permission.)

> know you're in it with all the others. You're going to attack me tomorrow – don't deny it, you're plotting on the telephone.' (Sutherland, 1976, p. 184)

This particular patient was relieved of his delusions by electroshock treatment, which has been extensively studied with animals.

The several kinds of schizophrenia may be sub-classes of a single homogeneous disorder, as set out in Figure 7.2, or they may represent several kinds of mental derangement that are only superficially similar. Simple schizophrenia is exemplified by lethargy and apathy. Such patients show no interest in the world and seem to live behind a plate of glass. They are colourless, strange and flat personalities who are often taken to be stupid.

Hebephrenia is the name sometimes given to a syndrome that includes silliness, seclusiveness and mannered or stereotyped behaviour. Such stereotypy, one of Bleuler's (1950) accessory symptoms of schizophrenia, is the basis of the amphetamine model of schizophrenia described below. Hebephrenics frequently talk to themselves, indulge in bizarre fantasies and suffer from hallucinations. Typically they show signs of strangeness from early childhood and are intensely religious. They may be taken for harmless cranks or they may be hospitalized if they are nuisances to their families.

Catatonia is a much more disturbing disorder. Such patients may remain motionless for days on end, holding any position into which their limbs are placed. The condition resembles tonic immobility in animals, which is sometimes taken to model it (see pages 184–5 below). During periods of excitement catatonics become aggressive, incontinent and sexually uninhibited. They are prone to imitate movements made by others, and in conversation to repeat whatever is said to them. Paranoia is sometimes taken as a form of schizophrenia and sometimes listed as a separate disorder. It includes many of the symptoms of hebephrenia with additional complications of delusions of grandeur or persecution. Chronic amphetamine intoxication also shares these symptoms (Connell, 1958), which is a link between human symptomatology and the amphetamine model of schizophrenia based on stereotypic movements.

In laboratory animals

Typically, the schizophrenic lives in a world of his own private fantasies, and as such has no obvious parallels in animals. However, just as there are humans who are somehow deficient in ordinary human qualities, so there are animals who clearly are not ordinary members of their species. It is a matter of investigation to see whether or not the deviant animals have common aetiologies with the deviant humans, and the extent to which the various human psychoses are the result of a common physical or psychological aberration.

It is unreasonable to expect animals to show psychoses of the paranoid type described above by Sutherland, because the manifestation of the disorder is the patient's self-description. A mute eavesdropper on a telephone conversation might be thought rude, brash or boorish, and he might seem phobic if he also showed more than an appropriate fear of the telephoner on other occasions, but he could not be called paranoid in the absence of a public misinterpretation of the telephone conversation. Evidence of this order of misinterpretation cannot be expected from animals. Nevertheless, schizophrenic reactions in animals are not unheard of, and a famous case is the dog, V3, studied by a team of researchers led by Gantt. Newton and Gantt (1968) quote a distinguished Scottish psychiatrist who saw the dog.

> As we would define schizophrenia in terms of disturbance of motility, sensation and overt behaviour, this is a schizophrenic dog showing stereotypy and ambivalence. The conclusion that the disorder from which this dog is suffering is akin to the human disease seems to my mind inescapable. If ever there was a schizophrenic dog, this is one. (Newton and Gantt, 1968, p. 51)

The overt behaviour of the dog that justified this diagnosis consisted of catatonic, cataleptic and stereotyped postures in which 'the dog would stand in a rigid posture, essentially immobile, for 5 to 30 minutes in the presence of persons, during which time his limbs could be flexed, extended, or put into all manner of bizarre positions in which they would remain for periods of time as long as 30 minutes'. On the basis of an abnormally high urine content of histidine exhibited in an analysis by paper chromatography, V3 was presumed to have a biochemical defect. Unfortunately, the dog was killed in a fight before a detailed biochemical examination could be carried out.

V3 was subjected to a variety of conditioning procedures associated wth experimental neurosis. Other aspects of psychotic-like behaviour appear in animals raised in social isolation or under unnatural restricted conditions. For example, Mitchell (1968) studied twelve three-and-a-half to four-and-a-half year old rhesus monkeys that had been raised for all or part of their first year in total isolation, and compared them with twenty other monkeys on the behaviour categories listed in Table 7.1. At one month of age the isolates were separated from their mothers and raised in enclosures that denied them any kind of social contact with other

Table 7.1 Definitions of clock-categories with the range of the between-tester reliability coefficients in parentheses

Behaviour	*Definition*
Fear–flight	Any fear grimace or running from another animal when not in a playful context. Often accompanied by screeching or submitting. Does not include non-social fear or fear of experimenter. (0.92–1.00)

Table 7.1 continued

Behaviour	*Definition*
Aggression–hostility	Barking and threatening accompanied by pilo-erection; pursuit of another animal while showing signs of rage; biting or tearing of fur or flesh. Does not include self-aggression or experimenter-directed aggression. (0.94 to 1.00)
Social play	Bouncing, running, wrestling, biting, and pulling without signs of hostility. Self-play is not included. (0.99–1.00)
Social sex	Socially directed mounting, thrusting, and presenting. Autocratic behavior is not included. (0.99–1.00)
Social explore	Any physical contact with another animal which cannot be scored as play, sex, fear, or aggression. (0.90)
Non-social play and exploration	All non-social locomotion except stereotypy. Manual exploration, oral exploration and self-play. (0.91)
Immobility	All non-social immobility except disturbance behavior. (0.95)
Stereotypy	Pacing and flipping. (0.99–1.00)
Disturbance	Rocking, crouching and wall-hugging. (0.99–1.00)
Locomotion	Walking, pacing, flipping, etc. All locomotion except self-play. (0.91–0.99)
Immobility	All immobility except disturbance. (0.95)
Coo	A soft drawn-out 'coo' sound occurring most often when the monkey is not active. (0.90–0.99)
Non-repetitive stereotypy	Weird, idiosyncratic, stereotyped movements which are easily recognized by their rigidity. Always absurdly out of context from the on-going behavior of the animal. (1.00)

The coefficient reported above is the gamma revision of Kendall's tau. From Mitchell, G. (1968) 'Persistent behavior pathology in rhesus monkeys following early social isolation', *Folia Primatologica*, 8, 132–47. Reprinted by permission of S. Karger AG, Basel.

animals and humans for either six or twelve months. Four of the isolates were kept in isolation from months 1 to 6, four from months 6 to 12, and four from months 1 to 12. The comparison animals were reared with companions either in the jungle or in the laboratory.

The animals were tested in three situations.

1 Stranger pairings in a familiar environment for 15 minutes per pairing.
2 Individual exposure to a novel environment for 15 minutes.
3 Dominance tests between isolate and non-isolate animals on eight occasions with each of a variety of partners.

Results of the stranger-pairing tests were that fear–flight, hostility and disturbance occurred more often in isolates than in controls, regardless of whether the isolates were paired with adults, age-mates or infants. In the novel environment, the isolates exhibited more signs of disturbance, more stereotyped behaviours and less 'coo' vocalizations (a common response to mother–infant separation in normal monkeys) than the non-isolate controls. Some details of the mean durations of these behaviours, shown by males and females separately, appear in Table 7.2. In the dominance tests there was no difference between isolate and control females, but isolate males fear-grimaced and retreated from control males significantly more often than the reverse.

Mitchell found stereotypies to occur more often and for longer periods in isolate than in control monkeys, and reports significant differences between the relative frequencies of repetitive and non-repetitive stereotypies according to whether the isolates were paired with adult or infant companions; repetitive pacing, flipping and body-rocking were significantly more frequent in the presence of adults, who also elicited a large amount of fear-grimacing by the isolates. Non-repetitive bizarre posturing appeared more frequently in the company of infants, when fear-grimacing was uncommon. Mitchell concludes:

> Isolate monkeys, particularly the 12-month and younger isolates, exhibit a high amount of repetitive stereotyped movements, such as rocking, an infantile form of behaviour which, like fear, wanes with age. Non-repetitive stereotyped movements, on the other hand, occur no more frequently in younger isolates than in older isolates . . . . These findings [in conjunction with others] show that repetitive movements indicate specific fear where non-repetitive movements ('catatonic contractures', 'ritualistic movements', 'abnormal limb posturing', 'floating-limb phenomenon', etc.) reflect the

Table 7.2 Mean durations of behaviours in the novel playroom with exact probability levels of the Mann-Whitney U test

Behavior	*Isolate male*	*Control male*	*P*	*Isolate female*	*Control female*	*P*
Disturbance	4.68	0.02	0.05	0.90	0.16	0.05
Non-repetitive stereotypy	0.18	0.00	0.03	0.34	0.00	
'Coo' vocalizations	0.02	0.15	0.02	0.11	0.53	
	4½-year-old isolates	3½-year-old isolates				
Play and explore	1.36	0.68	0.03			
Locomotion	4.82	1.97	0.04			

From Mitchell, G. (1968) 'Persistent behavior pathology in rhesus monkeys following early social isolation', *Folia Primatologica*, 8, 132–47. Reprinted by permission of S. Karger AG, Basel.

> presence of more deeply based disturbances in deprived primates. Perhaps they are symptoms of true 'psychosis'. (Mitchell, 1968, p. 145)

True psychosis and psychotic behaviour

True psychosis is a condition identifiable in humans; animals exhibit psychotic behaviour and may or may not suffer from true psychoses. True psychosis and psychotic behaviour can be distinguished according to their controlling variables. True psychosis is almost certainly a biochemical deficit that will eventually be corrected by pharmacological treatment. Psychotic behaviour is amenable to modification by reinforcement in much the same manner as non-psychotic behaviour, and can be so modified without affecting the true psychosis. For example, using cigarettes as reinforcers, Ayllon, Haughton and Hughes (1965) conditioned a hospitalized schizophrenic woman to carry a broom wherever she went in the hospital. Some unsuspecting psychiatrists offered various psychodynamic interpretations of this behaviour based on the nature of the woman's psychosis, but Ayllon and his colleagues were able to control the appearance of the symptom merely by the dispensation of the cigarettes. They were not able, though, to affect the woman's true psychosis. The same is true of token reinforcement and other behaviour modification systems; they control the frequency of psychotic behaviours without altering a patient's true psychosis if such a condition exists.

Nevertheless, even this much is a formidable achievement, for both practical and theoretical reasons. Practically, behaviour modification programmes establish prosthetic environments in which disabled patients may function in a reasonably normal manner (Ayllon and Azrin, 1968). Theoretically, they demonstrate that symptoms of psychosis are not directly determined by

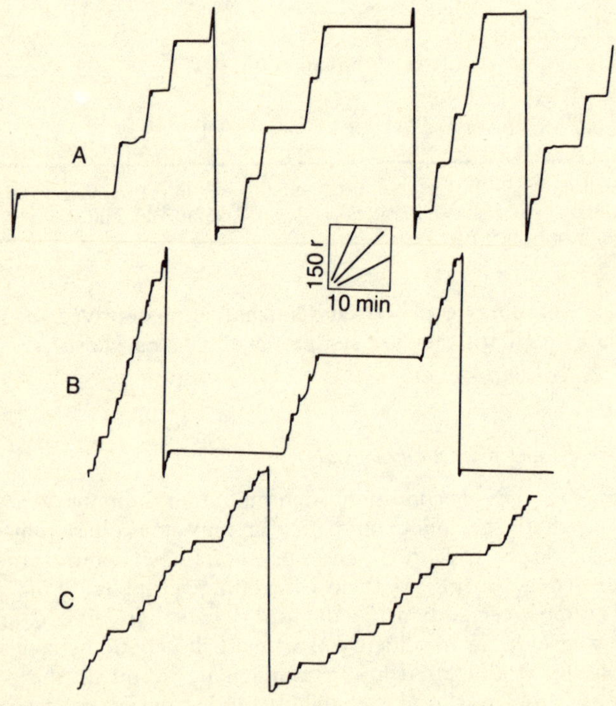

Figure 7.3 Cumulative response records of a pigeon (A) and two psychotic patients (B,C) where a large number of responses (Fixed Ratio) are required for reinforcement (food for the pigeon; trinkets for the patients). All the records illustrate 'ratio strain' shown by occasional long pauses before the bird or patient begins a ratio after reinforcement. (From Skinner, B.F. (1957) 'The experimental analysis of behavior', *American Scientist*, 45, 343–71, Reprinted by permission.)

whatever biochemical lesion occasions psychosis. True psychosis makes a person different from a normal human, but the form of the psychotic behaviour depends on environmental contingencies. Such a bifurcation of causal agents is implied in the clinical concepts of process and reactive schizophrenia, and of primary and secondary depression, both of which differentiate endogenous from exogenous psychosis.

Figure 7.3, from Skinner (1957), shows psychotic human and normal animal behaviour controlled by fixed-ratio schedules of reinforcement. Both the pigeon record (A) and those of the patients (B, C) display the usual run and break characteristics of behaviour reinforced on fixed-ratio schedules. The pigeon pecked a key for grain and the patients operated a plunger for candy, coins or cigarettes. The patients, both severely ill according to Skinner, behaved normally according to the schedule of reinforcement. This is not always so with psychotics, as is apparent from Figure 7.4. In this figure are cumulative records of plunger operation by one normal (A) and three psychotic (B,C,D) humans reinforced with nickels on a variable-interval 1-minute schedule. Record A is typical of behaviour rates generated by this type of reinforcement schedule; records B, C and D show sustained behaviour by the psychotics, but it appears in erratic fits and starts, rather than at a steady pace as is normal. The cumulative records in Figures 7.3 and 7.4 remind us that not all of the behaviour of human psychotics is psychotic behaviour, and that the behaviour of human psychotics that responds to environmental contingencies is not all of the behaviour of psychotic patients. An animal model of psychotic behaviour is easily arranged by proper control of environmental contingencies, as with the patient described above, but an animal model of true psychosis must be sought in another dominion.

Model psychoses

The psychedelic model of distorted perception

HUMAN AUTO-INTOXICATION Osmond and Smythies (1952) proposed that schizophrenia is a form of auto-intoxication, and along with Abram Hoffer (Hoffer, Osmond and Smythies, 1954) offered research evidence in support of this view. It was one of these workers who had the experience of being a sea anemone described on page 161. There is other evidence that biochemical factors play an important part in schizophrenia, although some of

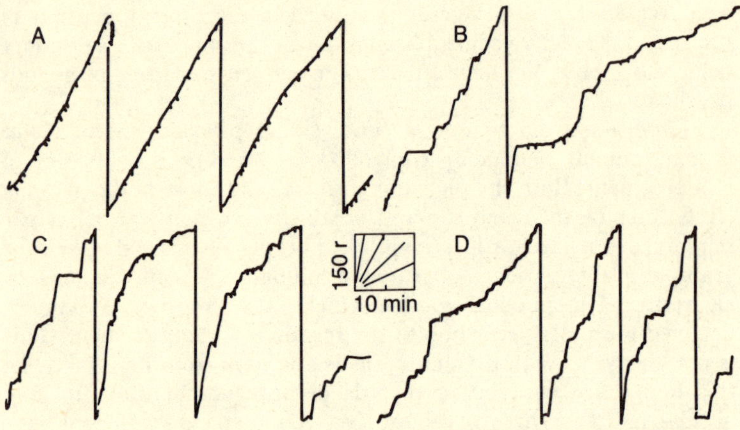

Figure 7.4 One normal (A) and three psychotic (B,C,D) performances on a variable interval reinforcement schedule with nickels as reinforcers secured on average once per minute. The normal record shows typical steady-state responding, but the patient behaviour is erratic and unpredictable. (From Skinner, B.F. (1957) 'The experimental analysis of behavior', *American Scientist*, 45, 343–71, Reprinted by permission.)

the early data are open to more mundane interpretations, like the quality of hospital food (Kety, 1959). Nevertheless, schizophrenic-like behaviour can be produced by psychotomimetic drugs, or hallucinogens, like mescaline and d-lysergic acid (LSD-25).

The trance-inducing effects of mescaline have been known to Western psychiatrists for over a hundred years but the psychotomimetic effects of d-lysergic acid were discovered by accident by the German researcher, Hofmann, in 1943, after working on the synthesis of the substance. As Hofmann (1959) describes it:

> In the afternoon of 16 April, 1943, when I was working on this problem, I was seized by a peculiar sensation of vertigo and restlessness. Objects, as well as the shape of my associates in the laboratory, appeared to undergo optical changes. I was unable to concentrate on my work. In a dream-like state I left for home where an irresistible urge to lie down overcame me. I drew the curtains and immediately fell into a peculiar state similar to drunkenness, characterized by an exaggerated imagination. With my eyes closed,

> fantastic pictures of extraordinary plasticity and intensive color seemed to surge towards me. After two hours this state gradually wore off. (Hofmann, 1959)

Since then, many others have described dramatic visual experiences induced by d-lysergic acid, including Aldous Huxley in *The Doors of Perception*. The drug is much more potent than mescaline and produces its hallucinogenic effects in minute doses. In recent years LSD-25 has become a cult drug and many similar compounds that produce psychotic-like states have been synthesized. Artificial psychosis produced by administration of these drugs has been called 'model psychosis' (McKellar, 1957) to indicate that the drug-induced condition may not be an exact replica of the actual pathological state. Modern belief is that model psychosis does not correspond to schizophrenia because perceptual distortions in schizophrenics are more typically auditory than visual.

DRUG-INDUCED ANIMAL BEHAVIOUR Present practice is not for psychiatrists and researchers to test the effects of drugs on their own feelings and perceptions so much as to examine how they affect standard animal behaviours. One such effort looked at the way LSD-25 and other drugs affect the web-building behaviour of spiders, and another has compared the effects of LSD-25 with doses of blood serum from schizophrenic patients on this behaviour (Bercel, 1960). Figure 7.5(a) is a photograph of a normal web of the *Aranea diademata* spider. Alongside it (Figure 7.5(b)) is an *Aranea diademata* web produced after 5 weeks of small doses (0.04 or 0.06 micrograms) of LSD-25 administered in sugar by mouth. According to Groh and Lemieux (1968) the centre of the web photographed in Figure 7.5(b) looks like a web built after administration of blood serum from a catatonic schizophrenic patient. Such a web is shown in Figure 7.6.

A less naturalistic study with LSD-25 was conducted with cats. Jacobs, Trulson and Stern (1976) gave twelve female cats intraperitoneal injections of either saline or LSD-25 in doses of 10.0, 25.0, 50.0 or 100.0 micrograms per kilogram, and had 'blind' observers rate three groups of the cats' behaviours. The first group (rubbing, treading, vocalizing) occurred infrequently in undrugged (zero dose) cats and was not affected by d-lysergic acid. The second group (staring, grooming, head or body shake) occurred frequently under control injections and even more frequently under the drug. The third group (limb flick, abortive grooming, investigatory and hallucinatory-like behaviour) hardly occurred at all except when the drug was injected.

At least four of the drug-affected behaviours – staring, limb

Figure 7.5 (a) Normal web of the Aranea Diademata spider.
(b) Web of the Aranea Diademata spider after a single small dose of LSD-25. The web becomes larger and more regular, except for the center which becomes irregular (see Figure 7.6). (From Groh, G., and Lemieux, M. (1968) 'The effect of LSD-25 on spider web formation', *International Journal of Addictions*, 3, 41–53, Reprinted by courtesy of Marcel Dekker, Inc.)

flick, abortive grooming, and hallucinatory-like behaviour – have psychotic qualities. Limb flicking suggests an unnatural sensation insofar as it is only performed by normal cats to shake dirt or water from a paw; also, like abortive grooming, it is a stereotypy of the kind already reported for primates in restricted environments. Finally, hallucinatory-like behaviour, 'scored when the cat looks around the floor, ceiling, or walls of the cage and appears to be tracking objects visually, or when the cat either hisses at, bats at, or pounces at unseen objects', is an obvious candidate for a psychotic interpretation.

Similar behaviour to this was observed by Mitchell (1953) in cats rescued from a flood and boarded in an animal clinic.

> All went well for five days . . . . On the sixth day most of them went off their food and mental symptoms began with one young tortoiseshell. She began chasing imaginary beings which seemed to appear to her like large flying insects. She tried to grab them in mid-air and cowered down as they zoomed towards her. Then one after another six of the 10 cats

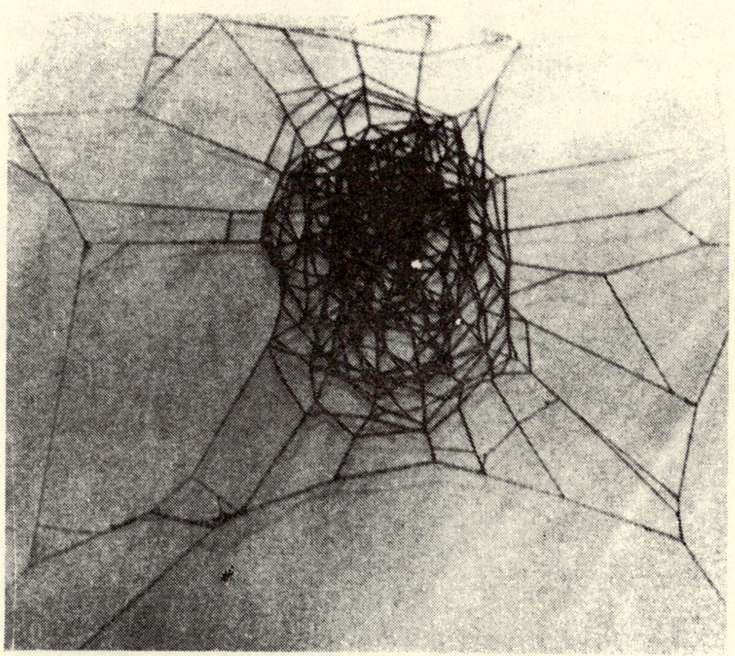

Figure 7.6 Web built by Aranea Diademata after absorption of blood serum from a catatonic schizophrenic patient. (From Groh, G., and Lemieux, M. (1968) 'The effect of LSD-25 on spider web formation', *International Journal of Addictions*, 3, 41–53, Reprinted by courtesy of Marcel Dekker, Inc.)

> took up the chase. . . . Three of the cats acted as though wading in water, picking their feet high and shaking them. . . . After four days the animals resumed eating and the mental symptoms gradually abated.

The similarity between the drug-induced and the apparently trauma-induced behaviour suggests that Mitchell's cats were probably affected more by diet than by trauma, a suggestion strengthened by the fact that the cats 'went off their food'. Diet-induced 'psychosis', caused by ergot infested rye, is a reputed disorder of humans at times of famine in the Middle Ages.

The objective of the above-mentioned drug studies was the isolation of the biochemical basis of psychosis. There are actually two methodologies in this enterprise. Kumar (1974) calls them the criterion drug and the criterion behaviour approaches. The first of these involves the classification of drugs according to their effects on behaviour in test situations; the second seeks animal analogues of human psychotic reactions. The criterion drug approach is the commercial method favoured for the evaluation of new antipsychotic agents, and is also a research method employed in the search for a common constituent of psychotomimetic compounds. It is basically an empirical approach insofar as it employs laboratory tasks that are affected by antipsychotic drugs without regard for their specific mode of action.

The criterion behaviour approach is more directly theoretical. Kumar (1974) lists four steps in its employment as a research strategy for discovering animal models of human aberrations.

1 The analysis of a human aberrant behaviour pattern for its basic characteristics.
2 The selection of aspects of animal behaviour as similar as possible to these characteristics.
3 A search for chemical agents that might affect these characteristics.
4 Tests of the effects of these agents on the human behaviour patterns originally analysed.

Steps in this direction are apparent with the amphetamine model of schizophrenia and the monoamine model of depression.

The amphetamine model of schizophrenia

With respect to the first two steps in Kumar's list, a basic characteristic of human schizophrenia that appears in lower animals is stereotypy. According to Bleuler (1950), these are outstanding among what he calls accessory (as against fundamental) symptoms of schizophrenia. Fundamental symptoms of human schizophrenic mentality, he thinks, are disordered associations and loss of emotion, which are internal private affairs. In addition:

> One of the most striking external manifestations of schizophrenia is the inclination to stereotypies. We find them in every sphere: that of movement, action, posture, speech, writing, drawing, in musical expressions, in the thinking and in the desires of the hallucinating patients. We find patients rubbing their right hands over their left thumbs intently and energetically for decades. Others will tap a saliva-moistened fingertip on every conceivable place. . . . Others beat on their beds rhythmically, clap their hands or perform all kinds of manipulations with their teeth, etc. (Bleuler, 1950, p. 185).

In comparison, the dog V3 (described on page 164) was diagnosed as schizophrenic on account of its catatonic, cataleptic and stereotyped behaviours, and stereotypies are prominent among the activities exhibited by other animals described as psychotic throughout this chapter. These stereotypies are occasioned by isolation from companions (deprivation stereotypies) and restriction of living space (cage stereotypies), but they can also be induced by drugs, particularly d-amphetamine. Thus, Randrup and Munkvad (1975) describe stereotyped 'looking at one forepaw, picking a certain part of the skin, seizing the bars of the cage, walking in fixed routes in the cage' as well as mouth movements and tongue protrusion, by monkeys injected subcutaneously with d-amphetamine sulphate.

With humans, amphetamine overdose reactions can be mistaken for symptoms of schizophrenia, a state of affairs that has stimulated the search for a model of schizophrenia via the induction and blockage of amphetamine reactions in animals. Studies in this direction constitute the third and fourth of Kumar's strategic steps inasmuch as they point to a rational basis for the action of antipsychotic drugs.

The drug treatment of psychosis is described by Silverstone and Turner (1982), where the clinical use of the drugs is evaluated. Here, the relevance of amphetamine psychosis to the relationship between true psychosis (as an emitted biochemical effect) and psychotic behaviour (as an elicited psychological effect) is considered. Clinically, amphetamine psychosis manifests itself by paranoia and delusions on the one hand, and by autonomous stereotypies on the other. It cannot be known if drugs of the amphetamine type, or of any other type, induce paranoia and delusions in animals, but many examples of drug-induced stereotypies in animals have been reported.

As well as examples from monkeys like the one described above, Randrup and Munkvad (1967) report repetitive stereotyped sniffing, licking and biting by rats and mice after acute d-amphetamine injections. Similarly, guinea-pigs and cats exhibited stereotyped head-movements; the guinea-pigs in the form of head-tossing, the cats in the form of sideways movements, as though looking around. Ellinwood, Sudilovsky and Nelson (1972) studied three classes of behaviour in cats under the influence of amphetamine; movement, posture and attitude. In the movement category they assessed various bodily sub-categories (head–neck, shoulder–foreleg, hip–hind leg, trunk and tail) according to their degree of coordination; under posture they included an index of ataxia (stagger, sway, wide-based legs) and an index of dystonic posture (disjunctive positions and movements); and under attitude they rated body-position in relation to the environment. The attitude category was divided into several modes of relating to the environment: unaware; indifferent; normal interest; abnormal interest ('restricted interest with a nervous, tense or apprehensive quality'); normal investigative; abnormal investigative ('actively reaching out to or approaching with compulsive character restricted elements of the environment'); reactive (nervous, jittery); normal focus; abnormal focus ('hooked on one aspect of environment with a tense and apprehensive quality').

Amphetamine produced abnormalities in all three behaviour classes; asynchronous movements, disjunctive postures, and compulsive attitudes, all of which can be seen as stereotyped repetitious behaviour at different levels of organization. Ellinwood *et al.* concluded that:

> Normally behavior is a symphony of changes and sequences of motivative behavior, all with their proper lability of attitude and spontaneous initiative for subsequent changes. There are proper relationships and smooth flow of control and autonomy not only in a spacial sense but also in a temporal sense. Following chronic amphetamine intoxication, components of behavior became relatively fixed over time and showed a loss of cohesive flow among different initiatives with their relative priorities. In addition, there was a segmentary autonomy manifested in the dyssynchrony of movements of separate body parts. In other words, there appeared to be islands of separate organization, each establishing its own autonomy or anarchy without integration into the larger behavioral symphony. (Ellinwood *et al.*, 1972, pp. 227–8)

If true psychosis is the biochemical lesion that de-synchronizes the larger behavioural symphony then repetitive stereotypies can represent the autonomy of lower levels of behavioural organization. They would be examples of psychotic behaviour. The question of whether such behaviour is responsive to the environmental circumstances of animals is answered affirmatively by Lyon and Robbins (1975) and Robbins (1976).

Robbins (1976) performed two experiments. In the first, twenty-eight thirsty male white rats were divided into two equal groups and trained to push a panel to reach water which was made available once every thirty seconds on the average. On each side of the panel was a bar, but these were inoperative at this stage of the experiment. For one group (CR), a stimulus light preceded each water presentation by 0.5 second; for the other group (FS), the same number of light presentations occurred at random. Thus, for group CR the light could become a secondary reinforcer (or discriminative stimulus) while for group FS there was no opportunity for the light to acquire special associative or eliciting properties. Then, half the animals in each group were injected with 10 mg/kg pipradol (a psychomotor stimulant of the amphetamine type) and all animals were tested for barpress acquisition with light onset as the reinforcer.

In contrast to the non-drugged subjects the drugged animals exhibited the expected repetitive head movements – sniffing and rearing stereotypies – but, whereas with the FS subgroup the stereotypies were aimless, with the CR subgroup they included a significant number of barpresses, indicating that drug-induced stereotypies are not impervious to their effects on the environment.

In the second experiment, Robbins examined whether the barpresses of drugged animals occur in spite of competing stereotypies or as part of the symphony of goal-directed behaviours that once secured water. This time, after injection, rats were required to press first one bar and then the other to secure the previously established secondary reinforcer. They did this without perseverating on the first bar, thus confirming the second of the alternative hypotheses; but they performed inefficiently with respect to the secondary reinforcer by perseverating on the second bar after reinforcement.

Robbins concluded that the 'results of both experiments support the hypothesis that a general action of stimulant drugs is to cause increased repetition of responding, with response selection being dependent in part on environmental contingencies', and compared the behaviour of amphetamine-treated animals with that of human schizophrenics. The concordance of amphetamine psychosis in humans with amphetamine-induced stereotypies in animals affirms the distinctions between true psychosis and psychotic behaviour, where the former is biochemically and the latter is situationally determined.

The monoamine model of endogenous depression

The simplest form of this model relates mood directly to monoamine neurotransmitter levels in the brain in unidimensional fashion; high serotonin, dopamine and noradrenaline underlie euphoria, and low levels accompany depression (Silverstone and Turner, 1982). The theory originated with the observation that *Raowolfia serpentina*, a herbal drug used in India for the treatment of hypertension, produced depressed mood as an undesirable side-effect. Reserpine, an alkaloid ingredient in the root of this plant, depletes monoamine storage in the brain, and the idea of a correlation between mood level and availability of these neurotransmitters arose. The model is supported by the actions of drugs used therapeutically for controlling depression.

Among such drugs are imipramine and iproniazid which, like *Raowolfia serpentina*, were originally used for other purposes and were found to have mood altering, this time antidepressant or euphoria-inducing, properties. These therapeutic drugs pertain to the monoamine model of depression via their pharmacological or neurotransmitter activity; imipramine inhibits re-uptake of both noradrenaline and serotonin, and iproniazid inhibits their deactivation. Thus, although their modes of action are different, each of these drugs increases the quantity of neurotransmitter available, and each moves mood from depression in the euphoria direction.

Ellison (1977) proposes a more complicated monoamine model of depression, than the unidimensional version. As summarized in Table 7.3, he noted that both serotonin (5-hydroxytryptamine, an indoleamine) and noradrenaline (a catecholamine) are depleted by reserpine (the double-depletion hypothesis) on the basis of which he devised experiments to deplete those neurotransmitters differentially. The procedure involved employment of the neurotoxins 6-hydroxydopamine and 5–6 hydroxytryptamine, which selectively deplete noradrenaline and serotonin storage, respectively. In low doses, these neurotoxins exert their action by occupying presynaptic sites of the respective neurotransmitters and so prevent them from reoccupying these sites during reuptake after stimulation.

With rats, Ellison reports effects of double depletion (norepinephrine and serotonin) and single depletion (norepinephrine or serotonin) on behaviour under positive and negative reinforcement situations, and on behaviour in an open-field arena. Response to the positive situation was measured by the reinforcing effects of sucrose or dim light onset on barpressing, while response to negative stimuli was assessed by speed of learning in a shuttlebox to avoid the onset of bright lights. With the open-field test, emotionality was assessed according to amounts of ambulation, exploration and grooming shown by the drug-treated animals in comparison with controls. With double depletion, Ellison found the rats to be under-responsive to positive reinforcement and over-responsive to the negative situation, paralleling the negative affect aspect of human depressive symptomatology. In the open-field arena the animals exhibited helpless-looking huddling.

With single depletion, low serotonin and low norepinephrine individually produced opposite effects relative to undrugged controls. With sucrose, low serotonin animals over-consumed while low norepinephrine animals under-consumed; and with bright light avoidance learning, low serotonin caused agitation and hyperactivity while low norepinephrine occasioned lethargy. In the open-field test the low serotonin group tended to huddle around the

Table 7.3 Summary of separate effects of serotonin and norepinephrine on: (a) behaviour maintained by positive and negative reinforcement; and (b) behaviour in the home colony or a novel open-field environment

(a) *Single depletion: positive* vs *negative reinforcement*

	Low serotonin	Low norepinephrine
Positive	Over-consumption of sucrose	Under-consumption of sucrose
Negative	Agitated	Lethargic

(b) *Single depletion: home* vs *novel environment*

	Low serotonin	Low norepinephrine
Home	Fearless	Fearful
Novel	Fearful	Fearless

edges, with much rearing but little general locomotion, whereas the low norepinephrine group were more active in the centre of the arena but showed little rearing or exploratory behaviour. Ellison's interpretation of these results is that the low serotonin group appeared to model neurotic anxiety, while the low norepinephrine animals looked more like models of depression.

In further experiments comparing the behaviours of single-depleted rats in familiar social and novel threatening environments, Ellison concluded that animals with high serotonin and low norepinephrine levels are driveless, withdrawn and depressive in their home colony environment but fearless and non-vigilant in the novel open-field. Low serotonin, high norepinephrine animals, on the other hand, were aroused and exploratory at home but frightened and paranoid in the open-field. Considering catecholamine (norepinephrine) levels alone, high and low levels accompany arousal and depression respectively in the home colony, but fear and non-vigilance, respectively, in the novel area. Thus a high catecholamine level appears to accompany either arousal or fearfulness according to environment familiarity or novelty, and a low level either occasions depression or fearless arousal, likewise depending on the environment. This view extends the simple unidimensional monoamine theory of mood that originally stimulated the reserpine model of depression, and suggests a bridge between this model and the more complex symptomatology of human manic-depressive psychosis.

Learned helplessness and exogenous depression

Seligman (1975) equates learned helplessness in animals with human depression on the basis of six common characteristics.

1 Lowered initiation of voluntary responses.
2 Negative cognitive set.
3 Dissipation over time.
4 Lowered aggression.
5 Loss of appetite.
6 Cholinergic overactivity.

Some modifications of the theory have recently occurred (Colotla, 1979) but Seligman offers a quantity of evidence from the animal literature in support of his learned helplessness model of depression. Here, 'negative cognitive set' may be considered as a fundamental experimental example. Seligman uses the concept to describe an experimental result in which animals first exposed to uncontrollable electric shock later find difficulty in learning a shock-avoidance response.

Seligman and Maier (1967) pre-trained three groups of dogs before submitting them to an escape learning procedure in a shuttlebox. In the pre-training, dogs in one group (escape) could terminate electric shocks to the hind legs by pushing a contact plate with their heads; dogs in a second group (yoked) received shocks exactly like the dogs in the escape group, but could not control them; and dogs in the third (naive) group were harnessed like the others but not shocked. The results of the second part of the experiment, in which all the dogs were trained to escape shock by shuttling from one to the other compartment of the shuttlebox, were that the dogs in the escape and naive groups learned to escape shock in the shuttlebox by jumping more and more rapidly trial after trial, but that no such learning occurred with the yoked group previously exposed to uncontrollable shock.

Seligman believes that the kind of helplessness exhibited by the yoked animals is parallel to the helplessness shown by human depressives. He attributes the helplessness to a learned 'negative cognitive set', but this interpretation is disputed. An alternative interpretation in terms of norepinephrine levels in the brain is proposed by Weiss, Glazer and Pohorecky (1976), whose experiment is described in Chapter 4. Whatever the explanation, the experimental result is not in doubt, and it is not unlikely that the origination of human depression lies in repeated failures similar to those experienced by the experimental dogs. This is not to say, though, that what the dogs feel parallels the feelings of depressed human patients, although the inference is possible on the basis of the criteria of animal suffering discussed in Chapter 2.

Disorders of mood and disorders of movement

Hyperactivity and elation

Robbins and Sahakian (1980) propose three characteristics – behavioural similarities, common aetiology and similar responses to treatment – as criteria for animal models of human disorders. On these accounts they describe animal models of mania in terms of hyperactivity, elation and irritability similarities, and of the effects of lithium on such behaviours in clinical and experimental situations. Hyperactivity, in animals, they suggest, might correspond to volubility, distractibility and flights of ideas in humans, although they recognize that the different ways of measuring hyperactivity in animals (running wheel, stabilimeter, jump–flinch, etc.) are not necessarily equivalent. Nevertheless lithium, an effective treatment for manic psychosis, attenuates a number of measures of activity in experimental animals.

As an operational measure of elation in animals, Robbins and Sahakian suggest a lowered threshold for positive reinforcement; that is, an animal may be said to be elated if it emits the same amount of behaviour for a lesser amount of reinforcement, or emits more behaviour for the same reinforcement magnitude. They offer two laboratory illustrations; intracranial self-stimulation and response to shifts in magnitude of reinforcement. For the first of these they choose the method of Stein and Ray (1960) for titrating self-stimulation current levels in rats. With this method, two bars are presented in an operant chamber equipped to deliver electrical stimulation to the brain. A press on bar 1 produces brain stimulation and simultaneously reduces the intensity of current available for the next response. A bar 2 press produces no additional stimulation but increases the current intensity that a bar 1 press produces. Thus shifts between bar 1 and bar 2 responses determine the level and frequency of intracranial stimulations, and Robbins and Sahakian suggest that elation is indicated when an animal selects a lower than normal intensity of stimulating current.

For their second analogue of elation Robbins and Sahakian invoke the so-called Crespi effect. This effect is that rats trained to traverse a runway with a given amount of reward run faster when the magnitude of reward is increased than do control rats trained with the higher reward magnitude from the beginning. Crespi's effect was shown with group comparisons; a similar phenomenon, called behavioural contrast, occurs with individual animals. In this case, pigeons shifted from simple interval schedules of reinforcement to schedules with interval and extinction components show increased rates of pecking in the interval component as pecking in the extinction component declines. Thus rate of response goes up in one component of the schedule as rate of reinforcement goes down in the other, which might be more an illustration of irritability than of elation. Ironically, behavioural contrast could be interpreted as schedule-induced aggression, which Robbins and Sahakian choose to operationalize as irritability in animals, but in this case they select pain-induced biting by monkeys, which may not be equivalent to the extinction-induced pecking by pigeons that relates to behavioural contrast.

Turning to drugs, Robbins and Sahakian review, among others, amphetamine and morphine models of mania. In suitable doses both of these drugs occasion hyperactivity that is blockable by lithium treatment. Amphetamine is also implicated in animal models of schizophrenia, but this is on account of the stereotypy induced by chronic high amplitude doses, whereas the hyperactivity-inducing property of amphetamine is more prominent at low doses. These dose-response effects of amphetamine were integrated by Lyon and Robbins (1975), who showed that in rats, increasing rates of response occur in a decreasing number of behavioural categories as amphetamine dose increases. Lithium treatment, Robbins and Sahakian conclude from several experimental reports, antagonizes amphetamine-induced locomotion but not amphetamine-induced stereotypic movements. Thus, if the low- and high-dose amphetamine models apply to mania and schizophrenia respectively, then lithium treatment with animals, as in the case of humans, applies specifically to mania but not to schizophrenia.

Acute morphine administration induces hyperactivity in cats, mice and rats, although the required effective doses range over a factor of more than a hundred (1.5 mg/kg with rats through 20 mg/kg with cats up to 225 mg/kg with mice.) In addition the hyperactivity takes different forms in the different species, such that cats claw and bite, rats rear and run in bursts, and mice show continual compulsive locomotion. In mice, lithium attenuates morphine-induced hyperactivity, but so do a number of other drugs not specific to the treatment of mania.

Hypoactivity and depression

In comparing animal models with their human originals, Ellison (1979) warns of difficulties in three areas: matching symptomatologies; comparing behavioural units; and the importance of social environments. Thus he observes that, with humans, symptomatology relates to verbal reports and diagnosis, while with animals it pertains to disturbances in basic behaviours; that, with humans, pathological units are expressed as classes of personal characteristics, while with animals they are specific responses to specific situations; and that while human psychopathologies occur in social contexts, animal models are normally based on non-social experimental situations. Ellison's own research (Ellison, 1977, 1979) has encompassed animals in actual social environments, but in other cases situational factors have been implicated in animal psychopathology by way of analysis of behavioural processes. Such a case is Ferster's (1966, 1974) account of depression.

Setting mood aside, Ferster notes that depression is inferred in organisms that exhibit lower than normal amounts of behaviour, and lists four kinds of situations in which the maintenance of normal behaviour is defective.

INSUFFICIENT BEHAVIOUR IS EMITTED FOR REINFORCEMENT In this case, the organism may occasionally emit normal behaviour but either infrequently or in insufficient amounts to secure reinforcers available in either the physical or the social environment. In the case of reinforcers available on interval schedules, they may be missed because behaviour is not emitted often enough to secure them; in the case of reinforcers available on ratio schedules, they may not be secured because insufficient responses are emitted to obtain them. In both cases, operant extinction occurs and behaviour that was weak before becomes even weaker and infrequent.

THE ENVIRONMENT IS AVERSIVE In this case the environment is deficient in positive reinforcers and overburdened with aversive events that generate withdrawal. Such a case could be loss of a job or a loved one and rebuff in the search of substitutes.

EMITTED BEHAVIOUR IS PUNISHED Punishment has several effects on behaviour, the most salient of which is to suppress it. Thus punished behaviour occurs with lower than normal frequency, and, if punishment is general and sustained, the suppression can affect a number of classes of behaviour in the manner of retarded depression. According to Seligman (1975) it can also lead to learned helplessness. However, in addition to suppression, punishment also induces aggression, so apart from depression a punishing environment can also occasion violent manic-type behaviour.

SUDDEN CHANGES IN THE ENVIRONMENT The environment in which an organism lives has both discriminative and reinforcing properties. Under normal circumstances the behaviour of an organism adjusts to these two stimulus properties, but if the environment undergoes sudden changes, the discriminative stimuli controlling the emission of a response may be degraded or omitted, so that the response no longer occurs; or if the response does occur, the environment may not provide the normal reinforcement. A dramatic laboratory example of this state of affairs is provided by the chimpanzee, Dennis, a subject in one of Ferster's experiments.

> Although the goal in these experiments was to build and then analyze a complex cognitive repertoire, we occasionally made errors, as any parent might, which weakened the overall repertoire of an animal seriously. One such incident occurred when we were forced to replace the female of the pair . . . . When we substituted another female the result was a profound disruption of the male's behaviour. He . . . bit her, pummelled her with fists, kicked her, or pushed her. He entered the work chambers fewer times than before and worked for shorter times whenever he did enter. For several weeks his food intake was less than 80 per cent of normal. The quality of his work, when it occurred, was not impaired. We could have described Dennis as angry and depressed, and we would not have been too far off the mark. (Ferster, 1966)

After such an environmental change, Dennis displayed an inadequate behavioural repertoire, and the new environment presumably supplied insufficient reinforcement. However depression is not the only result of this state of affairs, as Dennis's violence and the cases of Cupid and Azov described in earlier chapters illustrate.

The tonic immobility model of catatonia

A lizard, turned on its back and held for a few seconds, is likely to remain in this position for some minutes after it is released. The same is true of chickens, rabbits and several other animals (Chertok, 1968). The phenomenon has been known for centuries, and was originally seen as an animal analogue of hypnosis. This is now a discredited notion, and the phenomenon is now more widely referred to as tonic immobility.

Although the immobile animal appears to be helpless and defenseless, it is possible that the 'immobility reflex' deters an attacker from killing the immobile organism. Darwin called the immobility 'death feigning', and this view prevails in the predator-reaction theory of Ratner (1967), in which immobility is seen as the last of a sequence of defensive behaviours that progresses from freezing, fleeing and fighting to immobility as a natural predator approaches. Gallup (1974) has summarized evidence that tonic immobility is a response to fear, and Gallup and Maser (1977) propose it as a model of human psychotic catatonia. In support of the model they offer the following characteristics as common to catatonic schizophrenics and tonically immobile animals.

1 Rapid onset.
2 Parkinsonian-like tremors.
3 Muscular rigidity.
4 Hypnotic or stuporous gaze.
5 Mydriasis.
6 Exophthalmos.
7 Waxy flexibility.
8 No loss of consciousness.
9 Sudden termination.
10 Absence of vocal behaviour.
11 Agitated aggression and defense reactions after recovery.
12 Occasionally fatal.
13 Onset associated with emotional stress.

According to a combination of these correspondences and Ratner's predator-reaction theory, catatonic schizophrenia is a human psychotic response to a highly aversive situation, and is the evolutionary end-point of an adaptive survival response in lower animals.

8 Disorders of childhood

Separation syndromes: animal and human

Species similarities

Zoologists, ethologists, veterinarians and psychologists have repeatedly catalogued animal behavioural disorders caused by inadequate quartering, by accidents of nature, by abrupt changes in physical and social surroundings, and by loneliness. Such disorders have been extensively investigated in field and laboratory studies of isolation-rearing and mother–infant separation in several species of monkey, breeds of dogs, chimpanzees, and cattle (Stephens, 1974).

Some experimental results with monkeys are described in earlier chapters. In dogs, Scott, Stewart and DeGhett (1973) report a *separation reaction* (distress vocalization) in the beagle on removal from its mother before 8 weeks of age, a *kennel-dog* or *isolation syndrome* (withdrawn, fearful, untrainable) in Shetland sheepdogs raised in isolated kennels to the age of 5 months, and a *depressive syndrome* in a fox terrier that was dominated by its littermates. The fox terrier illustrates the sensitivity of the depressive syndrome to the social environment.

> Terriers reared [with littermates] frequently gang up on one member of the group and attack it persistently (Fuller, 1953). This particular animal had long been the object of such persecution. In the run, it stood listlessly, head and tail down, seldom ate, and appeared to be sick. We therefore transferred it to a litter of nonaggressive cocker spaniels in a nearby run. Within half an hour, it had recovered, holding its head and tail up, interacting constantly with the cockers, and eating. Left in the group of cockers, it continued to behave normally. (Scott, Stewart and DeGhett, 1973, p. 15)

Depression is also attributed to monkeys either reared in isolation or periodically separated in infancy from their peers and mothers. Kaufman and Rosenblum (1967a) report three such cases. Four infant pigtail monkeys lived in a group that included their mothers and father and a childless adult female. When the infants were between 4.8 and 6.1 months old they were separated from their mothers for a period of 4 weeks. The infants responded first by *agitation*, in the form of searching movements, distress cooing, and self-sucking and huddling; then, in three cases, after 24 to 36 hours, by *depression*, in which the infants withdrew and curled into a ball; and finally, by *recovery* after about six days. In all cases, after they were reunited with their mothers, the infants clung more to them than did normal infants of their age.

The depressive posture assumed by three of the four separated infants is shown in Figure 8.1, which also depicts a disinterested pigtail mother and her own child. The infant that did not show the depression component of the separation syndrome was the daughter of the dominant female in the group. While her mother was away this infant attached herself to other females of the group, which is uncharacteristic of pigtail monkeys but is common with bonnet monkeys.

The separation-depression posture can be controlled by the visual presence or absence of the mother. Kaufman (1973)

Figure 8.1 The monkey on the right illustrates the characteristic posture of an infant pigtail monkey separated from its mother. The left-hand mother-infant pair ignore it. See Figure 8.2 for a comparison with bonnet monkey behaviour. Redrawn after Kaufman, I.G. and Rosenblum, L.A. (1967b) 'The reactions to separation in infant monkeys: Anaclitic depression and conservation-withdrawal', *Psychosomatic Medicine*, 29, 648–75.

describes the effects on their infants when two pigtail mothers were simultaneously removed from a group. Infants of both mothers showed agitation, and both huddled into the separation-depression posture shown in Figure 8.1, but one case was mild and the other was severe. Both infants recovered, and then the mothers were returned to the group, but in cages so that they could be seen but not clutched. This state of affairs lasted 30 minutes, during which the infants vocalized and tried to reach their mothers. Both infants were unsuccessful, and the mildly depressed one soon resumed playing. The other infant remained huddled and immobile until its mother was taken out of sight. Kaufman (1973) says:

> On many days during the next 8 weeks, the mothers were each reintroduced for 30 minutes. Each time, the first infant after a cursory look continued what it was doing; but each time the second infant, after looking at its caged mother, collapsed into a ball and remained withdrawn and immobile until its mother was removed. (Kaufman, 1973, p. 50)

Kaufman interprets this behaviour as a helplessness response to the appearance of the mother. Another possibility is that the infant's behaviour is genetically transmitted as part of an et-epimeletic behavioural system that elicits attention and caregiving behaviour by the mother; the presence of the mother elicits huddling, and in the mother's absence huddling disappears.

Kaufman does not report whether this infant's depressive syndrome extinguished or strengthened with repeated re-introduction of the mother. In a similar case with a human child, the separation–withdrawal syndrome subsided after it had been repeatedly elicited with a 'stranger'. The child, Monica, was a relatively neglected infant who was hospitalized for surgery at the age of 15 months. At admission she 'had a wizened, drawn and dejected appearance' and was described as 'very depressed'. Physically she was underdeveloped and underweight. She remained in hospital, with occasional visits from her parents, for 9 months, and became something of a pet to the ward personnel. However in the presence of a stranger a depression–withdrawal reaction occurred that lasted as long as the stranger remained. (The context of the report suggests that the stranger would have been a white-coated investigator, but the description is not explicit.) When the stranger appeared:

> There was prompt loss of muscle tone . . . . Thereafter the baby manifested profound immobility, movements being restricted to glancing at the 'stranger' from time to time . . . [until she] gave this up and merely stared past him. . . . The

> empathic feeling of most observers was sadness, depression, hopelessness. (Engel and Reichsman, 1956, p. 434)

This reaction lasted only while the stranger was present and occurred only if he was passive. Any attempt to stimulate the child elicited retreat, whimpering and sobbing.

Concerning the overall sequence of the agitation, depression and recovery syndrome, Kaufman and Rosenblum (1967b) suggest an interpretation based on the adaptive value of each component of the syndrome. At first, they propose, separation from the mother induces noisy agitated behaviour that permits her to locate a missing infant; then, if the mother does not arrive, the infant huddles in a posture that both conserves body energy and reduces the chance of discovery by a predator; finally, the infant recovers and begins to move such that it satisfies its own needs independently of the mother.

The primate deprivation syndrome

Working with rhesus monkeys, Mason (1968) identifies four components of a primate deprivation syndrome.

1 Abnormal postures and movements, like huddling and rocking,
2 Motivational disturbances, like excessive fear of unfamiliar objects,
3 Poor integration of motor patterns of behaviour, like defective sexual performance,
4 Deficiencies in social communication, like inadequate submissive responses to threats by dominant troop members.

Mason explains this syndrome on the basis of a developmental trend in infant monkeys from a need to reduce to a need to raise the level of psychological arousal. He notes that the newborn monkey's behaviour is dominated by clinging, clutching, sucking and vocalizing, all directed to finding and maintaining contact with the mother. Normally this behaviour declines with increasing age but, according to Mason, the absence of the mother leads to abnormal rocking and self-clasping because the infant is forced to provide itself with stimulation normally provided by the mother as she carries and feeds the child. Stimulation of this sort is presumed to lead to arousal reduction.

As the infant gets older, exploration and play come to dominate its behaviour. Mason calls these behaviours 'arousal-inducing'; they involve the infant with novel events in the environment and in social interactions with its peers. Deprived infants, says Mason, lose the opportunity to adapt to environmental novelties and so react to them with excessively fearful emotion on meeting them when they are older. Defective sexual and communicative skills at maturity are thought to be the results of lost learning opportunities that would otherwise occur in play with age- and cage-mates during infancy.

Species differences

Mother–infant separation has different results according to age, sex, species, and the circumstances of separation. This last instance is illustrated by the case of the fox terrier described above, and also by the exceptional infant in the study of Kaufman and Rosenblum (1967a) who attached herself to substitute females. Such attachments exemplify the difference between bonnet and pigtail macaque monkeys in response to separation from their mothers alluded to above. Figure 8.1 above shows the depressive postures adopted by a macaque pigtail monkey separated from its mother, and also the indifference of another mother sitting with her infant. Figure 8.2, on the other hand, portrays the status of motherless bonnet monkeys. Such an infant is not abandoned to a life of loneliness but is comforted by other females who happen to be available.

Concerning age and sex, Hinde (1976) notes that, with rhesus monkeys, mother–infant interactions vary according to the infant's age, and that variations differ between the sexes. Up to 10 weeks of age, Hinde reports, males spend more time in contact with their mothers than do females, but after 30 weeks of age, females are on their mothers longer than are males. Thus, he suggests, different effects of mother–infant separation can be expected between the sexes, because the mother–child interaction is sex-dependent; and the sex difference could differ according to age of separation, because mother–child interactions with both sexes are age-dependent. However, as the fox terrier mentioned above demonstrates, there are individual differences within a sex.

Sex apart, in the matter of development and stereotypic movements there are recognizable species differences. Deprivation stereotypies increase in complexity from monkeys through chimpanzees to humans, and take on different species-characteristic forms (Berkson, 1967). Macaque monkeys rock back and forth while sitting, but chimpanzees rock and sway from side to side while either sitting or walking. Some human retardates rock while others twiddle their thumbs. Berkson refers to these respectively as 'rockers' and 'twiddlers'.

In dogs, Scott and Fuller (1965) emphasize age and species

Figure 8.2 A bonnet monkey mother with her own infant and two others that she carried and nursed through their periods of separation from their mothers. Redrawn after Kaufman, I.G. (1973) 'Mother-infant separation in monkeys: An experimental model', in J.P. Scott and E.C. Senay (eds), *Separation and Depression: Clinical and Research Aspects*, Washington: American Association for the Advancement of Science.

differences in the formation of mother–puppy emotional attachments. They differentiate four stages of canine development.

1 *Neonatal*, which lasts up to 2 weeks and is primarily devoted to securing nutrition by nursing,
2 *Transition*, which is a 'metamorphosis' from a puppy to an adult dog behaviour pattern that occurs in the third week after birth.
3 *Socialization*, through peer play and interaction from 3 weeks to 3 months of age.
4 *Juvenile* stage that lasts until the age of 6 months.

By this age the dog is fully adult and totally mobile. Separation of the dog from its mother in the first of these stages produces disturbances in the mother–puppy relationship and arouses the puppy to distress vocalization. Isolation from peers in the third stage, however, disturbs the establishment of dominance–submission social hierarchies through play. In this repect it is notable that the manifestation of dominance is species-specific. Scott and Fuller (1965) report that basenjis show dominance hierarchies over food but not over territory, but that with sheepdogs the situation is reversed; these dogs form dominance hierarchies with respect to space, but not with respect to food.

Differences between the reactions of Airedale terriers and German shepherds reared in kennels under conditions of partial isolation are described by Krushinskii (1962). Using withdrawal and submission (cowardice), which he calls the passive defensive reaction, Krushinskii (1962) reports that 9 of 111 (8.1 per cent) Airedales and 39 of 58 (67.2 per cent) German shepherds showed a strong or very strong expression of this reaction. On the other hand none of 41 Airedales and only 10 of 62 (16.1 per cent) German shepherds raised in individual homes showed strong submissive reactions.

Comparing humans and dogs, Scott and Fuller (1965) note differences in developmental characteristics at the time members of these species are normally weaned from their mothers. At that time, dogs are capable of adult eating and movement patterns so that socialization occurs through play with peers, not in a feeding relationship with the mother. Humans, on the other hand, learn to walk relatively later and are first socialized by the mother while she is still directly responsible for feeding. This difference would naturally affect separation syndromes shown by humans and by dogs to the extent that play and contact-seeking behaviours relate to feeding.

The human separation syndrome

In the case of humans, social services exist to replace missing mothers, but what is offered in physical comforts, formal education, and care and nourishment may not be sufficient to offset the loss of personal attention and love. Wayne Dennis (1973) found that although they were well taken care of physically, most of a group of girls orphaned in Lebanon grew up emotionally and intellectually handicapped. When adoption was possible, those taken into the care of private families by the time they were 2 years old seemed to recover completely from effects of the institutional experience. After that, recovery was partial at best. Human distress and disablement in adulthood and childhood are dependent on childhood experiences and can be minimized or prevented if adverse experiences are identified and controlled.

Ways of raising human children have particularly interested behaviourists and psychoanalysts; behaviourists because poor rearing practices cause disorders of learning, psychoanalysts because inadequate mothering causes disorders of personality and even death. If for nothing else, Freud would be remembered for his insistence on infantile sexuality, and Watson for his claim:

> Give me a dozen healthy infants, well formed, and my own specified world to bring them up in and I'll guarantee to take any one of them at random and train him to become any type of specialist I might select – doctor, lawyer, artist, merchant-chief and yes, even beggar-man and thief. . . . (Watson, 1924, p. 104)

Presumably, freaks, madmen, neurotics and emotionally disabled are also included, although Watson only went as far as to condition a mild fear of an animal in the well-known infant, Albert (Watson and Rayner, 1920). However others (Krasnogorski, 1925) reported experimental neurosis in children analogous to experimental neurosis in dogs soon after Watson's challenge appeared.

On the psychoanalytic side, Spitz (1945; Spitz and Wolf, 1946) and Bowlby (1969, 1973) describe separation syndromes in human infants. Spitz (1945) presents the original problem in graphic terms.

> At the beginning of our century one of the great foundling homes in Germany had a mortality rate of 71.5% of infants in the first year of life. In 1915 Chapin enumerated ten asylums in the . . . United States . . . in which the death rates of infants admitted during their first year of life varied from 31.7% to 75% by the end of their second year . . . . Dr. Knox of Baltimore stated that in the institutions of that city 90% of the infants died by the first year. (Spitz, 1945, pp. 53–4)

One paediatrician remarked on an infant mortality rate of nearly 100% at one institution in 1915 (Bakwin, 1942), and another is quoted as saying:

> I can given an instance from an institution that no longer exists in which, on account of the very considerable mortality among infants admitted, it was customary to enter the condition of every infant on the admission card as hopeless. That covered all subsequent happenings. (Bakwin and Bakwin, 1966, p. 444)

After better medical care was introduced in institutions for the care of orphans, mortality rates fell and exposed a psychiatric syndrome of anaclitic depression, characterized by sadness, weepiness, retarded development, stupor, loss of appetite and insomnia (Spitz and Wolf, 1946). Spitz reports finding nineteen severe and twenty-six mild cases of this syndrome in a group of 123 institutionalized human infants who were more or less competely separated from their mothers for a 3-month period at the age of 6 or 7 months. The infants became weepy, worried, slow-moving and insomnic, appearing physically as well as psychologically sick. When three mothers and children were re-united the children quickly recovered; otherwise, Spitz claims, physical and emotional development is permanently retarded. Some such infants die while others may exhibit a failure to thrive syndrome (FTTS) characterized by subnormal growth rate, unusual watchfulness, lack of cuddling, little smiling and subnormal vocalizations (Lipsitt, 1979).

One of the Spitz and Wolf cases that recovered is described as follows.

> This is a moderately intelligent, unusually beautiful child with enormously big blue eyes and golden curls. At the end of the eleventh month the child, who never had been very active, began to lose interest in playing . . . . She sat in a sort of daze, by the hour, staring silently into space . . . . If a toy was put into contact with her she would withdraw into the farthest corner of her bed and there sit wide-eyed, absent, and without contact, with an immobile rigid expression on her beautiful face. (Spitz and Wolf, 1946, p. 315)

This child was separated from her mother at the age of 7 or 8 months. At twelve months 'a certain measure' was taken and 'the syndrome disappeared'. The 'measure' was the restitution of the mother to the child.

With somewhat older children, in the 2- to 3-year-old age range, Bowlby describes a separation reaction of protest, despair and detachment. At first, the child cries and struggles to regain the mother (protest), then it becomes apathetic and withdrawn (evidence of despair) and finally it shows hostility and withdrawal (detachment) should the mother return after an absence of several weeks or months. The protest and despair stages are also shown by infant rhesus monkeys, but detachment seems specific to human infants.

Influenced by ethological methods and findings with animals, Bowlby (1976) now prefers a behavioural to a psychoanalytic interpretation of the protest–despair–detachment syndrome. He lists six ways in which animal studies can contribute to human problems. The first four are adopted from Hinde (1972), the last two Bowlby added specifically with the infant separation syndrome in mind.

1 By developing methods of observation, recording and analysis.
2 By the experimental study of analogous problems in animals in ways that would be unethical with humans.
3 By developing theoretical concepts.
4 By questioning extant theories of human problems.
5 By drawing attention to animal problems that may be typical problems of humans.
6 By providing confirmation of claims concerning human behaviour that are opposed for emotional reasons.

Hinde's four contributions to this list are pertinent to the value of animal studies in solving human problems in general. Bowlby's contributions relate to the value of findings with animals in sensitizing clinical investigators to a human problem in need of urgent attention, and *vice versa*. Just as abandoned human babies were once raised in orphanages that attended to the babies' physical requirements at the expense of their psychological necessities, so it has been standard laboratory practice to house small animals in individual home cages as a routine care and maintenance matter. The deleterious effect of this practice with infant monkeys is now well known, largely on account of pioneering work by H.F. Harlow. Since this original work, several studies on effects of isolation-rearing and infant–mother separation in rhesus monkeys have been conducted. These studies show similarities in symptomatology between human and monkey babies raised in isolation and contribute to the analysis of the origination of separation syndromes in human and other animal species.

Deprivation rearing: experimental procedures

About 1950, Harlow began an important line of investigation relevant to animal behavioural disorders. Unlike Pavlov, he did not set out to establish neurotic behaviours in animals, but instead he used them to study infant rearing practices that produce emotional disturbance in human children. Harlow originally studied the nature of love – not that of a mother for her mate, but the love of a child for its mother (1958).

In several experiments in Harlow's Wisconsin laboratory, rhesus monkeys reared with surrogate mothers were found to prefer cuddly cloth-covered models to wire models fitted for feeding with babies' bottles. The infant monkeys spent most of the day on the cloth models, and ran to them instead of to the feeding mothers whenever they were frightened. Infants temporarily removed from their real or surrogate mothers vocally protested in the same way that human children do in similar circumstances.

Since the initial preference studies, many rearing practices involving various degrees of social and perceptual isolation have produced unnatural or pathological behaviours in monkeys. Several methods have been described. They include the following.

Surrogate rearing: models

Infant monkeys are separated from their mothers and reared solely with surrogate models for several weeks. Mason (1968) used two kinds of models. They were both cloth-covered and fitted with feeding bottles, but one kind was fixed to the floor while the other was suspended from the roof of the cage. The first model is the usual stationary kind that the infant could clasp, but the second was free to swing. Mason reasoned that normal babies are rocked as well as cuddled by their mothers, and that babies brought up on the laboratory fixed models are deprived of rocking as well as of physical contact. Infants raised with fixed and swinging mothers in Mason's study were equally emotionally labile and socially inept on testing, but those with the swinging mothers did not engage in stereotyped rhythmic rocking, or take up the awkward abnormal postures that monkeys reared with fixed surrogates often adopt.

Surrogate rearing: humans

Infant monkeys are reared for their first months in nurseries with human 'mothers' substituting for monkey mothers as daytime nurses and attendants. These monkeys may later prefer human to monkey companions, but not to the extent of courting them, as Lorenz describes with imprinted birds. The monkeys' preferences are not as stable as imprinted choices. Preferences between monkeys and humans are restored to normal when the human-raised monkeys grow up with peers. They then choose the company of monkeys over humans.

Group rearing without parents

Infant monkeys are separated in infancy from their mothers, but not from their peers. This is called together–together rearing. Common to monkeys reared in this way is a group line-up called 'choo-choo', in which the dominant monkey sits at the front sucking its thumb and the others in the peer group sit in line behind him clasping each other chest to back.

Partial isolation

Partial isolation is where infant monkeys are raised for 6 months or more in cages from which they can see other monkeys but cannot touch them. Infants raised this way exhibit behaviours characteristic of autistic human children, like self-clasping, stereotyped back and forth rocking and self-mutilation. After long isolation, Cross and Harlow (1965) report signs of apparent dissociation in a monkey that seemed to lose control of a limb, and occasionally attacked it. This study is described more fully in Chapter 3.

Temporary separation

With this procedure, monkeys reared normally by their real mothers are separated from them for shorter or longer periods. Individual differences have been studied with this method. Kaufman and Rosenblum (1967) obtained signs of anaclitic depression in three out of four pigtail monkeys treated this way as described above. Hinde and Spencer-Booth (1970) studied individual differences in sixteen rhesus monkeys with this method. They separated the infants from their mothers for a 6-day period at ages varying from 25 to 32 weeks. They computed a 'distress index' based on the inverse of locomotor activity, the frequency of sitting and the rate of choo-calling, and found that the infants who showed the greatest disturbance after separation tended to be those who tried to maintain the most proximity to the mothers before separation.

Battering mothers

Infants are raised by mother surrogates that sometimes frighten, hurt or reject them. With 'rejecting' models that gave off air-blasts, vibrated violently, or that protruded spikes, when the infants climbed on them, Harlow and Harlow (1971) found that the infants clung more tenaciously than to the normal surrogate mothers. As the Harlow's say, 'Where else can the frightened baby go?' This method was deliberately used in an effort to create experimental neurosis in baby monkeys. It failed with the monkeys, but, the Harlows report, almost produced it in themselves.

Motherless mother rearing

In this case, infants are reared by their real mothers who themselves were separated from their own mothers and raised with surrogate mothers. Thus the babies are raised normally by their mothers, but the mothers' own upbringing was abnormal. The babies suffered. The mothers, who first had difficulty becoming mothers because of sexual incompetence, were also incompetent child-rearers. In contrast to feral monkey mothers, the motherless mothers were indifferent to their offspring when protection was needed, were cruel, and were inadequate concerning feeding. The offspring themselves are hyperaggressive, which is one of the few characteristics of impoverished rearing not overcome when infant monkeys subsequently mix with normal peers.

Total isolation

Infants are raised in total isolation with virtually no perceptual or social stimulation for several months. Monkeys reared like this suffer when paired with normal controls in being withdrawn, inept and apathetic, but the effect does not seem to be irreversible, at least with monkeys isolated for up to a year.

Solitary confinement

Animals reared under variations of any of the foregoing conditions are incarcerated in isolation pits for various periods of time. Harlow and Harlow (1971) took four 8-month-old infant monkeys that normally lived in pairs and placed them in solitary confinement pits for 2 days a week for 10 weeks. They report:

> During this period of time the animals showed progressive increases in self-clinging and clasping behaviors when out of the pits. Much to our surprise all showed self-aggression and two engaged in violent aggression against their partners . . . . All four monkeys showed disturbances, but there were marked individual differences in rates of induction. Two infants went into fits of autistic posturing in less than one month, a third started three weeks later, and the fourth showed little until after eight weeks of the pit treatment. (Harlow and Harlow, 1971, pp. 225–6)

With those methods, Harlow and his group were sometimes able to produce behavioural abnormalities in the monkey infants, and sometimes not. In this they paralleled the case with infant humans, who also respond in different ways when separation and sensory deprivation happen by circumstance or by chance.

Clinical syndromes in human children

Early infantile autism

Some of the symptoms of anaclitic depression, in human babies, and part of the separation syndrome in infant monkeys, resemble the early infantile autism syndrome in human children. The unresponsiveness and vacant staring of Monica (p. 188) and the child described by Spitz (p. 194) are common features of childhood autism, as are the rocking and huddling stereotypies shown by some animals on separation from their peers and mothers. A case of separation-depression in an infant monkey that huddled when its mother appeared, and then resumed normal playing when she left, was described earlier in this chapter. Similar behaviour in a female chimpanzee raised in a human home is reported by Temerlin (1975). He specifically compares the chimpanzee's behaviour to autism in human children.

> Lucy rocks when frustrated . . . she will sit still and rock back and forth, her movements regular and repetitive, looking neither to the right nor to the left, staring straight ahead. She at these times exhibits a kind of repetitive behavior often seen in autistic children . . . . While this behavior may have a self-stimulation function, *it elicits a quick comforting response from her mother*. . . . When Lucy is locked in her room she may protest and rock, *but the rocking disappears when Jane leaves*, only to return when Lucy hears Jane's footsteps approaching. (Temerlin, 1975, p. 38, italics added)

The development of Lucy's rocking is not reported, and the behaviour was not accompanied by other autistic symptoms in the chimpanzee. However, in human autistic children, rocking often occurs with self-destructive headbanging, and a case of this in a young rhesus monkey is described in detail by Levison (1970).

The monkey was one of six animals reared in a surrogate-mother chair that held the infant upright and provided a furry clutchable surface as well as a continuous supply of milk from an artificial nipple. During the first 100 days from birth, the monkey, normally in isolation, was handled daily in the course of pattern vision tests, and also during medical treatment for an injured arm. At 100 days of age he was transferred to an individual cage. He did not show the stereotyped rocking, crouching, self-biting and self-clutching usually exhibited by isolation-reared rhesus monkeys, probably on account of his unusual handling for medication. At one year of age the monkey was accidentally struck a glancing blow on the head by the door of a carrying cage, after which crouching, rocking and headbanging occurred in the presence of the cage. After this, headbanging was a regular occurrence whenever the monkey was made to wait while other monkeys were fed, as well as on other occasions that could be labelled as frustrating. However, as with Lucy, headbanging apparently only occurred when a person was in the room, or was about to feed the monkey.

Not all autistic posturing is under such clear stimulus control. The monkey infants subjected to periods of solitary confinement by Harlow and Harlow (1971) showed increasing amounts of autistic posturing with increasing periods of isolation experience, and they also became more aggressive when re-caged with a familiar peer. In these cases posturing and aggression increased with length of separation, but they did not elicit explicit tender loving care.

The syndrome of early infantile autism in human children is more complex than that which appears in monkeys. A check list of syndromes devised by Polan and Spencer (1959) divides the symptoms of autism into five major categories: language distortion; social withdrawal; activities lack integration; obsessiveness and nervousness; and family characteristics, of which the first and last obviously cannot apply to lower animals. The remaining three categories contain several behaviours that could be attributed to caged and isolation-reared monkeys, but the most salient instances appear in the category of obsessiveness and nervousness. In particular, these are rhythmic movements of the body, mechanical repetition of aimless movements and facial grimaces and twitchings. These are all more or less common behaviours found in isolation-reared animals.

An alternative animal model of autism is that of Tinbergen and Tinbergen (1972) who propose a theory of autism based on Tinbergen's (1953) motivational analysis of courting behaviour of the herring gull. This analysis postulates an approach–withdrawal conflict in the (timid) female as she approaches the initially aggressive male. Applied to human infants, the theory postulates 'that in each encounter with adults every infant is initially in a state of motivational conflict between social or bonding tendencies, and timidity or fear'. In normal children with normal parents the socially positive contacts overcome the waning fear, just as it does with the female herring gull, but with 'over-timid' children the fear is not overcome and 'bonding' does not occur. Instead, fear is elicited by the normally positive stimulus (e.g. a friendly smile) such that 'the child enters into a vicious circle, in which the overactivating of fear becomes progressive'. This state of affairs is the opposite of that sought by behaviour therapy with reciprocal inhibition. In that case, approach responses are strengthened while competing fear responses are deconditioned, which is the essence of the treatment of autistic children that the Tinbergens recommend on the basis of their ethological analysis.

Hyperkinesis

Hyperkinesis is a condition of human children that is characterized by restlessness, overactivity, short attention span, and poor coordination. At school the children are often discipline problems on account of their failure to concentrate, to pay attention, or to sit quietly when requested to do so by the teacher. Table 8.1 shows a list of teachers' complaints about such children to the staff of a psychiatric clinic (Conners, Eisenberg and Barcai, 1967). Hyperkinetic children have poor interpersonal relationships that can cause them to grow into delinquent and antisocial adults, although Cole and Moore (1976) attribute the delinquency more to the social situation than to the nature of the child itself. Minimal brain damage (MBD) is often suspected as a cause of the disorder, although strong evidence of neurological deficit is seldom evident and many authorities regard the MBD designation as more fictional than factual (Yule, 1978). An illustrative case is that of Mike, described by Wender (1971).

> Mike was referred to the clinic when he was 8 years old because of persistent behavioral and academic problems.

Table 8.1 Reasons for teachers' referrals for children with learning disabilities and school behavioural problems who as a group benefited from dextroamphetamine treatment in comparison to placebo

Hyperkinetic syndrome	*Non-hyperkinetic symptoms*
Hyperactive	Shy
Short attention span	Sullen
Restless	Slow, inactive
Fidgety	
Belligerent	
Impulsive	

Extracted from Conners, C.K., Eisenberg, L., and Barcai, A. (1967) 'Effects of dextroamphetamine on children. Studies on subjects with learning disabilities and school behavior problems', *Archives of General Psychiatry*, 17, 478–85.

> When he was 6, his foster mother characterized his behavioral problems as follows: 'At school his teacher tells me that he disrupts the class socially although he is very bright. Michael is still hyperactive. There is hardly a still moment in his day, but by 8.00 p.m.if he is put to bed, he goes to sleep quickly . . . . He cries a lot, especially when told he cannot do something, when told he must come in from play, and also at mealtimes, because he does not want to eat . . . . He does not play well with the neighbourhood children . . . he always wants to fight, wrestle, and poke people. . . . When he was 8 . . . a pediatrician . . . placed the boy on 5mg of d-amphetamine (long-acting).' A report from his teacher three months later read as follows: 'Mike has improved considerably . . . . He feels he's taking the "magic" pill and so he can do no wrong while under the influence of it. . . . His foster mother reported a marked improvement in behavior at home and an approximately 50% decrease in his neurosis.' (Wender, 1971, p. 198).

Two characteristics of Mike are his aggression and his positive response to treatment with amphetamine. The successful use of amphetamine in calming some hyperactive children was first reported by Bradley (1937). He treated thirty problem children between the ages of 5 and 14 years with benzedrine (d-l-amphetamine) and notes a 'spectacular' improvement in fourteen of them. Seven children experienced euphoria and reported feeling peppy, which is a normal response to amphetamine, but others who had been domineering and aggressive became placid and easy-going during amphetamine treatment. This paradoxical effect of amphetamine has been the basis of several proposals for an animal model of the hyperkinetic syndrome.

Animal models of hyperkinesis

Amphetamine response in the dog

One research group (Corson, Corson, Arnold and Knopp, 1976) investigated amphetamine with violent dogs. In classical conditioning studies of psychophysiological reactions to unavoidable electric shocks, this group examined the effects of amphetamine on eighteen dogs, eight of whom were untrainable and diagnosed as hyperkinetic. The dogs were housed in large (9 ft by 3 ft) air-conditioned runs and were periodically exercised and fed to maintain good health and normal weights. During baseline sessions, ten of the dogs stood quietly in the conditioning chamber for 2-hour periods while physiological measures were taken. The hyperkinetic dogs, however, bit and destroyed their harnesses in 15 to 20 minutes. After oral administration of amphetamine 1 hour before experimentation, five of the eight hyperkinetic dogs 'exhibited dramatic normalizing responses'. This proportion (62.8 per cent) is within the range of success with amphetamine-treated hyperkinetic children (Wender, 1971). Corson *et al.* report that two of their dogs became even more unmanageable than usual under amphetamine. This also occurs in 5 to 10 per cent of children diagnosed as hyperkinetic. Corson *et al.* (1976) claim that their untrainable dogs form a group of animals that has eight characteristics in common with those of hyperactive children.

1 The animals show uncontrollable hyperkinetic behaviour in a restraining situation.
2 The animals do not respond to reward or punishment.
3 Some animals are impulsive, intolerant of frustration, and cannot delay gratification.
4 Some animals are calmed by amphetamine treatment.
5 These animals are trainable and acquire conditioned responses.
6 Learned responses are retained after drug withdrawal.
7 The suppression of hyperkinetic violence continues after drug withdrawal.
8 The syndrome appears to be maturational and could have a genetic component.

On the basis of these they propose their animals as models of hyperkinetic children, although no specific aetiological analysis is presented.

Maturational lag in the rat

The maturational aspect of the hyperkinetic syndrome is the basis of an alternative animal model proposed by Campbell and Randall (1975). This models rests on relationships between age and activity levels shown by human children and laboratory rats, and on the age-related effects of amphetamine on these levels. Some authorities (cf. Wender, 1971) have compared the hyperkinetic syndrome to a 'maturational lag', as though the hyperactive patient, who often outgrows the syndrome at puberty, passes through a phase of heightened activity after such a phase would normally have ended. With human children, this is at the age of 5 years.

Rats also exhibit a period of hyperactivity in infancy. With them, general activity levels begin to rise at about 10 days of age, reach a peak at 15 days, and decline to adult levels between 5 and 10 days later. Campbell and Randall hypothesized that, just as amphetamine tranquilizes the hyperactive child, so it could reduce activity levels of infant rats. They measured the locomotor activity of 15-day-old and 30-day-old rats (eight per group), either isolated or with an anaesthetized adult female, in an 8 × 16 inch polyethylene cage under 0.0, 0.5 and 1.0 mg/kg amphetamine doses. The time course of the drug effect under each condition was examined, and four effects were discovered.

1 When isolated, in the undrugged state (saline injection) the younger animals were more active than the older rats.
2 In isolation, amphetamine raised activity levels in dose-related fashion, especially for the younger rats.
3 The presence of the anaesthetized adult had no effect on the older animals.
4 In the presence of the anaesthetized adult, activity levels of the younger rats were minimal at all dose levels.

This fourth result resembles the calming effect of amphetamine on hyperkinetic children. However, with infant rats, it was not that they were calmed by amphetamine, but that the presence of the adult female prevented the hyperactivity that amphetamine otherwise induces. This is because the infants remained by the anaesthetized female throughout experimental sessions regardless of drug dose, whereas the older animals were not in contact with her for most of the time.

Mason (1970) reports a similar result with chimpanzees to that of Campbell and Randall. He found that young chimpanzees allowed to choose between 'roughhouse' playing or clinging to a stimulus-person normally spend more time playing than clinging in the undrugged state, but that this choice is reversed following amphetamine administration, to the extent that roughhousing is almost nonexistent about 100 minutes after injection. This result is also similar to the calming effect of amphetamine on hyperkinetic children, and Mason explains it on the basis of generalized arousal.

The neonatal split-brain kitten

Sechzer (1977) reports that the paradoxical effects of amphetamine appear in neonatal split-brain kittens. Her group sectioned the corpus callosum and psalterium of thirteen kittens at either 1 or 3 days of age and compared ten of them with five normal littermates on wholebody activity and discrimination learning at the age of 6 months and 12 months, respectively. Without amphetamine the activity of a group of six of the lesioned animals, subjectively assessed according to locomotion, ease of handling, exploration and alertness, was higher than that of the five normal controls, but as d-amphetamine doses were increased up to 1.0 mg per kilogram of bodyweight the assessed activity of the split-brain group decreased dramatically, while that of the controls increased. Unfortunately the ratings were not made blindly, inter-rater reliabilities are not reported, and the reasons for including only a sub-group of the lesioned animals in the comparison are not given. Nevertheless, similar results were reported with an open-field apparatus in which the number of squares crossed in a 3-minute test were objectively counted.

In the discrimination task, conducted when the animals were 12 months old, they were reduced to 80 per cent of their normal bodyweights and trained on a simultaneous visual discrimination task to open one of a pair of doors at the end of an alleyway where small pieces of raw spleen were located. Some cats were trained first with amphetamine and then without it while others were trained first without and then with the drug. In both cases, Sechzer reports, the split-brain animals performed more poorly than the unoperated controls, but improved appreciably in the drugged condition. Without amphetamine, she says, the split-brain cats were easily distracted and worked through thirty trials a day only with difficulty, but completed this number of trials easily and quickly when amphetamine was administered. Sechzer lists a number of similarities between hyperactive children and neonatal split-brain kittens, including motor activity, attention, learning and memory, emotionality, impulse control, coordination and response to amphetamine. She does not claim that hyperactive human children are literally split-brained, but speculates about an impairment in norepinephrine or dopamine release at neural junctions.

Miscellaneous models

Hyperactivity can result from many factors, including housing in isolation, undernourishment, anatomical and chemical lesions, and lead poisoning (Goldberg and Silbergeld, 1977), but of these only hyperactivity due to lead poisoning parallels childhood hyperkinesia with respect to the calming effect of amphetamine. Michaelson, Bornschein, Loch and Rafales (1977), however, show that this effect could be the result of undernourishment, not of lead poisoning directly. These data were obtained with mice. With dogs, Bareggi, Becker, Ginsburg and Genovese (1979) suggest a possible genetic basis of hyperactivity on account of differences in the abilities of purebred beagles and Telomian dogs to hold a sitting position on command. The Telomians (Malaysian dogs from the Telom river area) were less capable of sitting still than were the beagles, and among hybrids some animals were more able to sit still after amphetamine treatment than were others.

However the childhood hyperkinetic syndrome does not consist only of hypermotility; maladaptive social behaviour, distractability and learning and memory deficits are also included. The effect of amphetamine, therefore, may not be a paradoxical calming of affected children but a normal alerting of children, making them more attentive to the task in hand. Lyon and Nielson (1979) adopt such a position in the case of simple repetitive problems.

They studied effects of methylphenidate on performance of repetitive and sequential tasks by monkeys, and showed that the drug improved performance on the former but adversely affected sequential task performance. This result is in accordance with the 'fragmented action' theory of amphetamine effects on behaviour put forward by Lyon and Robbins (1975).

The different reactions of different children to amphetamine treatment may ultimately prove more interesting as a diagnostic than as a therapeutic tool inasmuch as the hyperkinetic syndrome, like schizophrenia, may comprise not one but a multiple of disorders. The critical problem then becomes that of matching the specific disorder of the child with his or her specific response to amphetamine doses, rather than that of exploring an animal model of a total hyperkinetic syndrome. On the other hand, if hyperkinesis is not a characteristic of a child but is a situationally-determined disorder, then analysis of the situations that elicit hyperactivity in animals can suggest ways of treating the condition when it appears in human children.

9 Summary and conclusions

> At present our animal models are all that we have to find new treatments for our mentally ill. After all, much of the progress of modern medicine came about in our lifetime because researchers were able to reproduce human disease in animal models and hence seek new ways of treatment. (Domino, 1976)

Animal models may be all that we have for finding new ways to treat the mentally ill, but treatment is not the only contribution that animal models make to clinical psychology and psychiatry. I have given examples of contributions to treatment, both treatment with drugs for psychosis and behavioural treatments for neurotic and psychophysiological disorders, but in addition I have shown how animal models have contributed to problems of theory, diagnosis and prevention. The theoretical contribution is the most fundamental of these, because it guides the search for practical applications; the humoral theory of Hippocrates and Galen guided treatment up to and beyond the madness business of George III, until the theory was displaced by modern discoveries in medicine. Late among these discoveries came the pharmacological treatments for psychosis, which are themselves the cornerstones for advances in theories of depression and the schizophrenias.

There is no longer any doubt that psychosis is a disorder of the brain and central nervous system, and it cannot be long before studies with animal models make the exact nature of this disorder known. This will not mean, however, that psychosis will then be reclassified from a functional to an organic disorder; the behavioural expression of an abnormal brain is not isomorphic with the disordered status of that brain. Animal studies have shown that consequences are powerful determinants of behaviour, and it is in this sense that psychotic behaviours are functional, not organic, abnormalities. The diagnostic consequence of this discovery is the opportunity to remove the confusion between psychosis as cause and psychosis as symptom. Lists and arrangements of psychotic symptoms are standard contents of psychiatric textbooks, and many of them are repeated in sections of this volume. These symptoms are frequently presented as signs of psychotic mental illness, as though the illness were their origin or cause, but the animal models show that the origins of psychotic symptoms are in defective neurochemical systems (a modern version of disproportions of humours!) but that the forms that they take are governed at base by classical and operant conditioning principles. The animals tell us that some mental illnesses are illnesses in the way that stomach pains are illnesses – mental suffering caused by a physical attack – and that others are more like tuberculosis, a public hazard that gives the possessor no pain. Endogenous depression is in the former category, schizophrenic reaction in the latter. Anxiety and hysteria stand in similar relation to each other, except that they are entirely behaviourogenic.

As for prevention, animal models have contributed most to paediatrics, relapse into drugs, readdiction and patient care in hospitals. In the last century survival rates of orphaned human infants were often less than one in two, and the situation until recently was almost as bad with laboratory-raised infant monkeys. Now, since psychiatrists and primatologists have joined forces, not only are deprivation syndromes recognized in many species as well as humans, but preventive measures in psychiatry, psychology and paediatrics have become standard practice. More than that, research continues with young 'monkey psychiatrists' for isolation stress reactions in older monkeys, and with pet therapy for the aged and emotionally upset human children.

With alcoholism and drug addiction, advances through animal studies have occurred in at least two major areas – animal models of self-administration, and animal models of relapse and readdiction. Of these, the immediate salience of the self-administration type lies in its similarity to the normal form of human drug consumption, but the deeper value of the model is its relevance for conditioned tolerance and withdrawal. Relapse after detoxification is almost total in the principal forms of human drug addiction, and the root of such readdiction is most likely conditioned effects of withdrawal to the drug consumption situation. Animal models of readdiction by way of withdrawal symptoms conditioned to self-administration already indicate how to prevent readdiction of detoxified humans.

The third contribution of animal models to prevention in psychiatry is the non-medical care of patients in institutions. Numerous literate individuals have recorded their sufferings in hospitals for mental disorders without notable effect in practice. Beyond these, hospital staff come to accept the stereotyped pacing, apathy and territoriality of patients as inevitable manifestations of their illnesses, and so beyond nursing control. Sometimes this may be the case, but the analogous behaviours of animals confined in sensory barren environments suggest the iatrogenic rather than pathogenic nature of these activities, and hence that patient suffering can be reduced by appropriate hospital reform.

A final question remains to be answered about the future. What kind of approach to animal models for psychiatry should be taken? Three kinds of methods are available, one used by Ellison (1979), one proposed by Hinde (1962), and one that I adopted in earlier parts of this book.

For Ellison, the question is one of criteria. What criteria of a human condition must an animal model meet? For schizophrenia, he recommends four.

1 *Face validity*: the animal model must be based on unique features of the psychopathology and its origination in the human case.
2 *Behavioural similarities*: animal and human psychopathologies must have common behavioural features.
3 *Biochemical similarities*: animals and humans must exhibit common biochemical and pharmacological features.
4 *Predictive value for therapeutics*: the animal model must contribute to therapeutic methods for humans.

For Hinde, the question is one of relevance. What animal behavioural processes are relevant for animal models? For neurosis, he recommends three.

1 Immediate animal responses to stress.
2 Ways in which these responses become habitual and distorted.
3 How these responses generalize to novel situations.

For myself, the question is one of behaviour theory. What are the laws of animal behaviour, and how might they apply to humans? Unlike Ellison, I do not seek ways of creating animal psychopathologies that resemble those of humans. Unlike Hinde, I do not seek equivalences between normal animal behavioural processes and abnormal behaviours of humans. Instead, I believe that abnormal behaviour is unexpected behaviour from the standpoint of normal theory; it is behaviour that proves the rules and opens the way to improve them. Whatever therapeutic gains accrue on the way, then so much the better for suffering humans, and for suffering animals as well.

In concluding, let me repeat a passage from the opening chapter:

> Recent studies of experimental neurosis from Russian origins . . . focus more on specific psychopathological disorders like gastric achylia, hyperglycaemia, sexual dysfunction, hypertension and cardiac insufficiency than on general neurotic emotionality, and this focus is generally true whenever modern animal analogues of human disabilities are sought. Thus not schizophrenia but the biochemical roots of stereotypy is the basis of some animal models of psychosis; not alcoholism but . . . animal and human responses to alcohol; not anaclitic depression but animal and human reactions to isolation rearing, are the foci of contemporary research on animal models in psychopathology.

It may be advisable, however, to regard this research as having less to do with models than it has to do with simulations. Research with models is a relatively recent enterprise for psychiatry, but for psychology, models have been employed for decades, and in science and engineering for centuries. Over this time, models have followed two principal courses; as small-scale versions of actual events and as simplified tests of functional relations. It is this second course that animal models for psychiatry are following. In no way are behavioural disorders in animals miniature versions of mental diseases in humans; if they are anything they are simulations of human conditions, exemplars of the stimulus control of behaviour. We have seen it argued (in Chapter 2) that animal models cannot, in fact, be anything because humans are uniquely different from animals; we saw also the argument that animal experiments *should* not be anything because animals and humans are so much alike. Overlooking the contradiction between them, I would answer the first of these arguments with the admission that it is *everything*, not anything, that animal models cannot be; and the second with the admonition that there is no choice. To the arguments combined I submit that it is only because we are different from animals that we care about their welfare: it is not the hungry Bengal tiger, who is troubled by animal experimentation.

References

Chapter 1 Psychopathology: the status of animals

AMERICAN PUBLIC HEALTH ASSOCIATION (1967) 'Vivisection–vivistudy: the facts and the benefits to animal and human health', *American Journal of Public Health*, 57, 1597–626.

AUER, E.T., and SMITH, K.U. (1940) 'Characteristics of epileptoid convulsive reactions produced in rats by auditory stimulation', *Journal of Comparative Psychology*, 30, 255–9.

AYLLON, T. (1979), 'Comments on "Two classic behaviour modification patients: a decade later" ', *Canadian Journal of Behavioural Science*, 6, 428.

BATESON, G. (ed.) (1961) *Perceval's Narrative: A Patient's Account of his Psychosis, 1830–1832*, Stanford, Stanford University Press.

BOWLBY, J. (1976) 'Human personality development in an ethological light', in G. Serban and A. Kling (eds), *Animal Models in Human Psychobiology*, New York, Plenum.

BRION, A. (1964) 'Les tics chez les animaux', in A. Brion and H.Ey (eds), *Psychiatrie Animale*, Paris, Desclée de Brouwer.

BROADHURST, P.L. (1960) 'Abnormal animal behaviour', in H.J. Eysenck (ed.), *Handbook of Abnormal Psychology*, London, Pitman, pp. 726–63.

BROADHURST, P.L. (1973) 'Animal studies bearing on abnormal behaviour', in H.J. Eysenck (ed.), *Handbook of Abnormal Psychology*, 2nd edition, London, Pitman, pp. 721–54.

CARRINGTON, R. (1959) *Elephants*, New York, Basic Books.

CHAMOVE, A.S., EYSENCK, H.J., and HARLOW, H.F. (1972) 'Personality in monkeys: factor analyses of rhesus social behaviour', *Quarterly Journal of Experimental Psychology*, 24, 496–504.

CHERTOK, L., and FONTAINE, M. (1963) 'Psychosomatics in veterinary medicine', *Journal of Psychosomatic Research*, 7, 229–35.

CHRISTIAN, J.J., and RATCLIFFE, H.L. (1952) 'Shock disease in captive wild mammals', *American Journal of Pathology*, 28, 725–37.

CRAWFORD, F.T., and PRESTRUDE, A.M. (eds) (1977) 'Animal hypnosis', *Psychological Record*, 27, special issue.

ELLEN, P. (1956) 'The compulsive nature of abnormal fixations', *Journal of Comparative and Physiological Psychology*, 49, 309–317.

ELLENBERGER, H.F. (1960) 'Zoological garden and mental hospital', *Canadian Psychiatric Association Journal*, 5, 136–49.

ELLISON, G.D. (1979) 'Animal models of psychopathology: studies in naturalistic colony environments', in J.D. Keehn (ed), *Psychopathology in Animals: Research and Clinical Implications*, New York, Academic Press, pp. 81–101.

EVANS, E.P. (1906), *Criminal Prosecution and Capital Punishment of Animals*, London, Heinemann.

EWING, L.S. (1967) 'Fighting and death from stress in a cockroach', *Science*, 155, 1035–6.

EY, H. (1964) 'Le concept de "Psychiatrie animale" (Difficultés et interet de sa problematique)', in A. Brion and H. Ey (eds), *Psychiatrie Animale*, Paris, Desclée de Brouwer, pp. 11–40.

FALK, J.L., SAMSON, H.H., and WINGER, G. (1972) 'Behavioral maintenance of high concentrations of blood ethanol and physical dependence in the rat', *Science*, 1977, 811–13.

FERGUSON, W. (1968) 'Abnormal behavior in domestic birds', in M.W. Fox (ed.), *Abnormal Behavior in Animals*, Philadelphia, Saunders, pp. 188–207.

FERSTER, C.B. (1966) 'Animal behavior and mental illness', *Psychological Record*, 16, 345–56.

FOX, M.W. (ed.) (1968) *Abnormal Behavior in Animals*, Philadelphia, Saunders.

FRAME, J. (1961) *Faces in the Water*, New York, Brazillier.

FRANK, G. (1975) *Psychiatric Diagnosis: A Review of Research*, Oxford, Pergamon.

FRASER, A.F. (1957) 'Intromission phobia in the bull', *Veterinary Record*, 69, 621–3.

FRASER, A.F. (1960) 'Spontaneously occurring forms of tonic immobility in farm animals', *Canadian Journal of Comparative Medicine*, 24, 330–3.

GAITO, J. (1976) 'The kindling effect as a model of epilepsy', *Psychological Bulletin*, 83, 1097–109.

GANTT, W.H. (1944) *Experimental Basis for Neurotic Behavior: Origin and Development of Artificially Produced Disturbances of Behavior in Dogs*, New York, Hoeber.

GAY, W.I. (1967) 'Comparative medicine', *Science*, 158, 1220–37.

GILBERT, R.M. (1977) 'Schedule-induced self-administration of drugs', in D.E. Blackman and D.J. Sanger (eds), *Contemporary Research in Behavioral Pharmacology*, New York, Plenum, pp. 289–323.

HINDE, R.A. (1962) 'The relevance of animal studies to human neurotic disorders', in D. Richter, J.M. Tanner, L. Taylor and D.L. Zangwill (eds), *Aspects of Psychiatric Research*, New York, Oxford University Press, pp. 240–61.

HOTHERSALL, D., and TUBER, D.S. (1979) 'Fears in companion dogs: characteristics and treatment', in J.D. Keehn (ed.), *Psychopathology in Animals: Research and Clinical Implications*, New York, Academic Press, pp. 239–55.

HUNT, H.F. (1964) 'Problems in the interpretation of "experimental neurosis" ', *Psychological Reports*, 15, 27–35.

JONES, B.E., and PRADA, J.A. (1973) 'Relapse to morphine use in dog', *Psychopharmacologia*, 30, 1–12.

JONES, I.H. (1971) 'Ethology and psychiatry', *Australia and New Zealand Journal of Psychiatry*, 5, 258–63.

JONES, I.H., and BARRACLOUGH, G.M. (1978) 'Auto-mutilation in animals and its relevance to self-injury in man', *Acta Psychiatrica Scandinavica*, 58, 40–7.

KEEHN, J.D. (ed.) (1979) *Psychopathology in Animals: Research and Clinical Implications*, New York, Academic Press.

KENDELL, R.E. (1975) *The Role of Diagnosis in Psychiatry*, Oxford, Blackwell.

KILGOUR, R. (1978) 'The application of animal behavior and the humane care of farm animals', *Journal of Animal Science*, 46, 1478–86.

KILLAM, K.F., KILLAM, E.K., and NAQUET, R. (1967) 'An animal model of light sensitive epilepsy', *Electroencephalography and Clinical Neurophysiology*, 22, 497–513.

KRUSHINSKII, L.V. (1962) *Animal Behavior – Its Normal and Abnormal Development*, New York, Consultants Bureau.

LEADER, R.W. (1967) 'The kinship of animal and human diseases', *Scientific American*, 216, 110–16.

LEADER, R.W. (1969) 'Discovery and exploitation of animal model diseases', *Federation Proceedings*, 28, 1804–9.

LEVY, D.M. (1952) 'Animal psychology in its relation to psychiatry', in F. Alexander and H. Ross (eds), *Dynamic Psychiatry*, Chicago, University of Chicago Press, pp. 483–507.

LIDDELL, H.S. (1944) 'Conditioned reflex method and experimental neurosis', in J.McV. Hunt (ed.), *Personality and the Behavior Disorders*, New York, Ronald, pp.389–412.

LIDDELL, H.S. (1956) *Emotional Hazards in Animals and Man*, Springfield, Thomas.

LINDSLEY, O.R. (1964a) 'Geriatric behavioral prosthetics', in R. Kastenbaum (ed.), *New Thoughts on Old Age*, New York, Springer.

LINDSLEY, O.R. (1964b) 'Direct measurement and prosthesis of retarded behavior', *Journal of Education*, 147, 62–81.

MASER, J.D., and SELIGMAN, M.E.P. (eds) (1977) *Psychopathology: Experimental Models*, San Francisco, Freeman.

MASON, W.A. (1968) 'Scope and potential of primate research', in J.H. Masserman (ed.), *Animal and Human*, New York, Grune and Stratton, pp. 101–18 (*Science and Psychoanalysis*, vol. 12).

MASON, W.A. (1976) 'Environmental models and mental modes. Representational processes in the great apes and man', *American Psychologist*, 31, 284–94.

MASSERMAN, J.H. (1943) *Behavior and Neurosis: An Experimental Psychoanalytical Approach to Psychobiologic Principles*, Chicago, University of Chicago Press.

McCANN, M.B., and STARE, F.J. (1967) 'The contributions of animal studies to nutritional discoveries that have benefited both animals and man', *American Journal of Public Health*, 57, 1597–604.

MENNINGER, K., MAYMAN, M., and PRUYSER, P. (1963) *The Vital Balance*, New York, Viking.

MEYER-HOLZAPFEL, M. (1968) 'Abnormal behavior in zoo animals', in M.W. Fox (ed.), *Abnormal Behavior in Animals*, Philadelphia, Saunders, pp. 476–503.

MIMINOSHVILI, D.I. (1960) 'Experimental neurosis in monkeys', in I.A. Utkin (ed.), *Theoretical and Practical Problems of Medicine and Biology in Experiments on Monkeys*, Oxford, Pergamon, (trans. R. Schacter).

MORRIS, D. (1955) 'The causation of pseudofemale and pseudomale behaviour: a further comment', *Behaviour*, 8, 46–56.

OTTENBERG, P., STEIN, M., LEWIS, J., and HAMILTON, C. (1958) 'Learned asthma in the guinea pig', *Psychosomatic Medicine*, 20, 395–400.

PAPPWORTH, H.M. (1967) *Human Guinea Pigs: Experimentation on Man*, London, Routledge & Kegan Paul.

PAVLOV, I.P. (1927) *Conditioned Reflexes*, Oxford, Oxford University Press.

RICHTER, C.P. (1957) 'On the phenomenon of sudden death in animals and man', *Psychosomatic Medicine*, 19, 191–8.

ROBINSON, H.J. (1967) 'Animal experimentation leading to the development of drugs benefiting human beings and animals', *American Journal of Public Health*, 57, 1613–20.

ROSENHAN, D.L. (1973) 'On being sane in insane places', *Science*, 179, 250–8.

SANGER, V.L., and HAMDY, A.H. (1962) 'A strange fright–flight behavior pattern (hysteria) in hens', *Journal of the American Veterinary Medical Association*, 140, 455–9.

SCOTT, J.P., and SENAY, E.C. (eds) (1973) *Separation and Depression: Clinical and Research Aspects*, Washington, American Association for the Advancement of Science.

SELIGMAN, M.E.P. (1975) *Helplessness*, San Francisco, Freeman.

SHAPIRO, M.B. (1975) 'The requirements and implications of a systematic science of psychopathology', *Bulletin of the British Psychological Society*, 28, 149–55.

SHERWOOD, G.G., and GRAY, J.E. (1974) 'Two "classic" behaviour modification patients: a decade later', *Canadian Journal of Behavioural Science*, 6, 420–7.

SIEGLER, M., and OSMOND, H. (1974) *Models of Madness, Models of Medicine*, New York, Macmillan.

STARTSEV, V.G. (1976) *Primate Models of Human Neurogenic Disorders*, Hillsdale, N.J., Erlbaum (trans. Marienne Schweinler and Vadim Pahn).

STOUT, C., and SNYDER, R.L. (1969) 'Ulcerative colitis-like lesion in Siamang gibbons', *Gastroenterology*, 57, 256–61.

TARSY, D., and BALDESSARINI, R.J. (1976) 'The tardive dyskinesia syndrome', in H.L. Klawans (ed.), *Clinical Neuropharmacology*, Volume 1, New York, Raven, pp. 29–61.

TINBERGEN, N. (1974) 'Ethology and stress diseases', *Science*, 185, 20–7.

TUBER, D.S., HOTHERSALL, D., and VOITH, V.L. (1974) 'Animal clinical psychology: a modest proposal', *American Psychologist*, 29, 762–6.

ULLMAN, L.P., and KRASNER, L. (1969) *A Psychological Approach to Abnormal Behavior*, Englewood Cliffs, Prentice-Hall.

VAN LAWICK-GOODALL, J. (1971) *In the Shadow of Man*, Boston, Houghton Mifflin.

WEISS, B., and SANTELLI, S. (1978) 'Dyskinesias evoked in monkeys by weekly administration of haloperidol', *Science*, 200, 799–801.

WINTERSCHEID, L.C. (1967) 'Animal experimentation leading to the development of surgical techniques', *American Journal of Public Health*, 57, 1604–12.

WOLPE, J. (1958) *Psychotherapy by Reciprocal Inhibition*, Stanford, Stanford University Press.

WOODRUFF, R.A., GOODWIN, D.W., and GUZE, S.B. (1974) *Psychiatric Diagnosis*, New York, Oxford University Press.

YEROFEEVA (EROFEEW), M.N. (1916) 'Contribution to the study of destructive conditioned reflexes', *Comptes Rendu de la Societé Biologique*, 79, 239–40 (abstract).

ZUBIN, J., and HUNT, H.F. (eds) (1967) *Comparative Psychopathology: Animal and Human*, New York, Grune and Stratton.

Chapter 2 Animal experiments and animal welfare

ARDREY, R. (1978) *The Territorial Imperative: A Personal Inquiry into the Animal Origins of Property and Nations*, New York, Bantam.

BANNISTER, D. (1981) 'The fallacy of animal experimentation in psychology', in D. Sperlinger (ed.), *Animals in Research: New Perspectives in Animal Experimentation*, New York, Wiley, pp. 307–17.

BASSFORD, H.A. (1981) 'The moral role differentiation of experimental psychologists', in J.D. Keehn (ed.), *The Ethics of Psychological Research*, Oxford, Pergamon, pp.27–31.

CHAPMAN, A.J., and JONES, D.M. (eds) (1980) *Models of Man*, Leicester, British Psychological Society.

COHEN, E.A. (1953) *Human Behavior in the Concentration Camp*, New York, Norton.

COLYER, R.J. (1970) 'Tail biting in pigs', *Agriculture*, 77, 215–18.

DARWIN, C.R. (1872) *The Expression of the Emotions in Man and Animals*, London, Murray.

DAVIS, H. (1981) 'Ethical considerations in the aversive control of behaviour', in J.D. Keehn (ed.), *The Ethics of Psychological Research*, Oxford, Pergamon, pp.61–7.

DAVIS, H., and WRIGHT, J. (1979) 'Procedural and parametric variability in studies of conditioned suppression', *Bulletin of the Psychonomic Society*, 14, 179.

DAWKINS, M. (1980) *Animal Suffering: The Science of Animal Welfare*, London, Chapman & Hall.

DIAMOND, C. (1981) 'Experimenting on animals: a problem in ethics', in D.J. Sperlinger (ed.), *Animals in Research*, New York, Wiley, pp.337–62.

EVANS, E.P. (1898) *Evolutional Ethics and Animal Psychology*, New York, Appleton.

FERSTER, C.B. (1966) 'Animal behavior and mental illness', *Psychological Record*, 16, 345–56.

FREY, S. (1976) 'The assessment of similarity', in M. von Cranach (ed.), *Methods of Inference from Animal to Human*, The Hague, Mouton, pp. 7–23.

GRIFFIN, D.R. (1976) *The Question of Animal Awareness: Evolutionary Continuity of Mental Experience*, New York, Rockefeller University Press.

HANSEN, R.S. (1976) 'Nervousness and hysteria in mature female chickens', *Poultry Science*, 55, 531–43.

HARLOW, H.F., GLUCK, J.P., and SUOMI, S.J. (1972) 'Generalization of behavioral data between nonhuman and human animals', *American Psychologist*, 27, 709–16.

HOTHERSALL, D., and TUBER, D.S. (1979) 'Fears in companion dogs:

characteristics and treatment', in J.D. Keehn (ed.), *Psychopathology in Animals: Research and Clinical Implications*, New York, Academic Press, pp. 239–55.

HUME, C.W. (1957) 'The strategy and tactics of experimentation', *Lancet*, 2, 1049–59.

KATZ, D. (1937) *Animals and Men: Studies in Comparative Psychology*, London, Longmans, Green.

KEEHN, J.D. (1967) 'Running and bar pressing as avoidance responses', *Psychological Reports*, 20, 591–602.

KEEHN, J.D. (1982) 'To do or not to do: dimensions of value and morality in experiments with animal and human subjects', in J.D. Keehn (ed.), *The Ethics of Psychological Research*, Oxford, Pergamon, pp. 81–4.

KILEY-WORTHINGTON, M. (1977) *Behavioural Problems of Farm Animals*, Stocksfield, Northumberland, Oriel Press.

LANE-PETTER, W. (1976) 'The ethics of animal experimentation', *Journal of Medical Ethics*, 2, 118-26.

LEVIS, D.J. (1979) 'The infrahuman avoidance model of symptom maintenance and implosive therapy', in J.D. Keehn (ed.), *Psychopathology in Animals: Research and Clinical Implications*, New York, Academic Press, pp. 256–77.

LEVISON, C.A. (1970) 'The development of head-banging in a young rhesus monkey', *American Journal of Mental Deficiency*, 75, 323–8.

LORENZ, K. (1966) *On Aggression*, London, Methuen.

MACDONALD, D., and DAWKINS, M. (1981) 'Ethology – the science and the tool', in D. Sperlinger (ed.), *Animals in Research: New Perspectives in Animal Experimentation*, New York, Wiley, pp.203–23.

MARCUSE, F.L., and PEAR, J.J. (1979) 'Ethics and animal experimentation: personal views', in J.D. Keehn (ed.), *Psychopathology in Animals: Research and Clinical Implications*, New York, Academic Press, pp. 305–29.

MASON, W.A. (1968) 'Early social deprivation in the non-human primates: implications for human behavior', in D.C. Glass (ed.), *Environmental Influences*, New York, Russell Sage Foundation, pp. 71–100.

MORRIS, D. (1967) *The Naked Ape*, London, Cape.

PAPPWORTH, W.M. (1967) *Human Guinea Pigs: Experimentation On Man*, London, Routledge & Kegan Paul.

REED, G. (1981) 'On being moral in immoral places', in J.D. Keehn (ed.), *The Ethics of Psychological Research*, Oxford, Pergamon, pp. 19–26.

ROSS, M.W. (1978) 'The ethics of animal experimentation: control in practice', *Australian Psychologist*, 13, 375–8.

ROSS, M.W. (1981) 'The ethics of experiments on higher animals', in J.D. Keehn (ed.), *The Ethics of Psychological Research*, Oxford, Pergamon, pp. 51–60. 51–60.

RUSSELL, W.M.S., and BURCH, R.L. (1959) *The Principles of Humane Experimental Technique*, London, Methuen.

SECHZER, J.A. (1981) 'Historical issues concerning animal experimentation in the United States', in J.D. Keehn (ed.), *The Ethics of Psychological Research*, Oxford, Pergamon.

SELIGMAN, M.E.P. (1975) *Helplessness*, San Francisco, Freeman.

SINGER, P. (1975) *Animal Liberation: A New Ethics for our Treatment of Animals*, New York, New York Review of Books.

SMYTH, D.H. (1978) *Alternatives to Animal Experiments*, London, Scolar Press.

STARTSEV, V.G. (1976) *Primate Models of Human Neurogenic Disorders*, Hillsdale, N.J., Erlbaum, (trans. Marienne Schweinler and Vadim Pahn).

STOUT, C., and SNYDER, R.L. (1969) 'Ulcerative colitis-like lesion in Siamang gibbons', *Gastroenterology*, 57, 256–61.

THORPE, W.H. (1974) *Animal Nature and Human Nature*, London, Methuen.

TINKLEPAUGH, O.L. (1928) 'The self-mutilation of a male macacos rhesus monkey', *Journal of Mammology*, 9, 293–300.

WALKER, S. (1982) *Animal Thought*, London, Routledge & Kegan Paul.

Chapter 3 Abnormal movements and convulsions

AUER, E.T., and SMITH, K.U. (1940) 'Characteristics of epileptoid convulsive reactions produced in rats by auditory stimulation', *Journal of Comparative Psychology*, 30, 255–9.

BARNETT, S.A. (1963) *The Rat: A Study in Behavior*, Chicago, Aldine.

BEACH, F.A. and JAYNES, J. (1954) 'Effects of early experience upon the behavior of animals', *Psychological Bulletin*, 51, 239–63.

BERKSON, G. (1967) 'Abnormal stereotyped motor acts', in J. Zubin and H.F. Hunt (eds), *Comparative Psychopathology: Animal and Human*, New York Grune & Stratton, pp. 76–94.

BERKSON, G., MASON, W.A. and SAXON, S.V. (1963) 'Situations and stimulus effects on stereotyped behaviors of chimpanzees', *Journal of Comparative and Physiological Psychology*, 56, 786–92.

BLEULER, E. (1950) *Dementia Praecox or the Group of Schizophrenias*, New York, International Universities Press.

CALHOUN, J.R. (1967) 'Ecological factors in the development of behavioral anomalies', in J. Zubin and H.F. Hunt (eds), *Comparative Psychopathology: Animal and Human*, New York, Grune & Stratton, pp.1–51.

CONNELL, P.H. (1958) *Amphetamine Psychosis*, Maudsley monograph no. 5, Oxford, Oxford University Press.

CROSS, H.A., and HARLOW, H.F. (1965) 'Prolonged and progressive effects of partial isolation on the behavior of macaque monkeys', *Journal of Experimental Research on Personality*, 1, 39–49.

DAVENPORT, R.K., and MENZEL, E.W. (1963) 'Stereotyped behavior of the infant chimpanzee', *Archives of General Psychiatry*, 8, 99–104.

DRAPER, W.A., and BERNSTEIN, I.S. (1963) 'Stereotyped behavior and cage size', *Perceptual and Motor Skills*, 16, 231–4.

ELLINWOOD, E.H., and KILBEY, M.M. (1975) 'Amphetamine stereotypy: the influence of environmental factors and prepotent behavioral patterns in its topography and development', *Biological Psychiatry*, 10, 3–16.

ELLINWOOD, E.H., and KILBEY, M.M. (1977) 'Chronic stimulant intoxication models of psychosis', in I. Hanin and E. Usdin (eds), *Animal Models in Psychiatry and Neurology*, Oxford, Pergamon, pp. 61–74.

ELLINWOOD, E.H., SUDILOVSKY, A., and NELSON, L. (1972) 'Behavioral analysis of chronic amphetamine intoxication', *Biological Psychiatry*, 4, 215-30.

EPSTEIN, L.H., DOKE, L.A., SAJWAJ, T.E., SORRELL, S., and RIMMER, B. (1974) 'Generality and side effects of overcorrection', *Journal of Applied Behavior Analysis*, 7, 385–90.

FALK, J.L., SAMSON, H.H., and WINGER, G. (1972) 'Behavioral maintenance of high concentrations of blood ethanol and physical dependence in the rat', *Science*, 177, 811–13.

FERSTER, C.B., and SKINNER, B.F. (1957) *Schedules of Reinforcement*, New York, Appleton-Century-Crofts.

FINDLEY, J.D., and BRADY, J.V. (1965) 'Facilitation of large ratio performance by use of conditioned reinforcement', *Journal of the Experimental Analysis of Behavior*, 8, 125-9.

FINGER, F.W. (1944) 'Experimental behavior disorders in the rat', in J.McV. Hunt (ed.), *Personality and the Behavior Disorders, vol. II*, New York, Ronald, pp. 413–30.

FINGER, F.W., and SCHLOSBERG, H. (1941) 'The effect of audiogenic seizures on general activity of the white rat', *American Journal of Psychology*, 54, 518–27.

FULLER, J.L. (1979) 'Genetic analysis of deviant behavior', in J.D. Keehn (ed.), *Psychopathology in Animals: Research and Clinical Implications*, New York, Academic Press, pp. 61–79.

GAITO, J. (1974) 'The kindling effect', *Physiological Psychology*, 2, 45–50.

GAITO, J. (1976) 'The kindling effect as a model of epilepsy', *Psychological Bulletin*, 83, 1097–109.

GAITO, J. (1979) 'The kindling effect: an experimental model of epilepsy?' in J.D. Keehn (ed.), *Psychopathology in Animals: Research and Clinical Implications*, New York, Academic Press, pp. 169–95.

GASTAUT, H., and TASSINARI, C.A. (1966) 'Triggering mechanisms in epilepsy. The electroclinical point of view', *Epilepsia*, 7, 85–138.

HINDE, R.A. (1962) 'The relevance of animal studies to human neurotic disorders', in D. Richter, J.M. Tanner, L. Tayler and O.L. Zangwill (eds), *Aspects of Psychiatric Research*, New York, Oxford University Press, pp. 240–61.

KANNER, L. (1943) 'Autistic disturbances of affective contact', *Nervous Child*, 2, 217–50.

KEIPER, R.R. (1970) 'Studies of stereotypy functions in the canary (*Serinus canarius*)', *Animal Behaviour*, 18, 353–7.

KILEY-WORTHINGTON, M. (1977) *Behavioural Problems of Farm Animals*, Stocksfield, Northumberland, Oriel.

KOKKINIDIS, L., and ANISMAN, H. (1980) 'Amphetamine models of paranoid schizophrenia: an overview and elaboration of animal experimentation', *Psychological Bulletin*, 88, 551–79.

KRUSHINSKII, L.V. (1962) *Animal Behavior: Its Normal and Abnormal Development*, New York, Consultants' Bureau.

LENNOX, W.G. (1941) 'Alcohol and epilepsy', *Quarterly Journal of Studies on Alcohol*, 2, 1–11.

LIBERSON, W.T. (1967) 'Withdrawal and fixation reactions in rodents', in J. Zubin and H.F. Hunt (eds), *Comparative Psychopathology: Animals and Human*, New York, Grune & Stratton, pp. 120–57.

LIDDELL, H.S. (1944) 'Conditioned reflex method and experimental neurosis', in J.McV. Hunt (ed.), *Personality and the Behavior Disorders*, New York, Ronald, pp.389–412.

LOVAAS, O.I., SCHAEFFER, B., and SIMMONS, J.Q. (1965) 'Building social behavior in autistic children by use of electric shock', *Journal of Experimental Research on Personality*, 1, 99–109.

LUISELLI, J.K. (1975) 'The effects of multiple contingencies on the rocking behavior of a retarded child', *Psychological Record*, 25, 559–65.

LYON, M. and NIELSON, E.B. (1979) 'Psychosis and drug-induced stereotypies', in J.D. Keehn (ed.), *Psychopathology in Animals: Research and Clinical Implications*, New York, Academic Press, pp. 103–42.

LYON, M. and ROBBINS, T.W. (1975) 'The action of central nervous system stimulant drugs: a general theory concerning amphetamine effects' in W. Essman and L. Valzelli (eds), *Current Developments in Psychopharmacology, Vol. 2*, New York, Spectrum, pp. 89–163.

McQUARRIE, D.G., and FINGL, E. (1958) 'Effects of single doses and chronic administration of ethanol on experimental seizures in mice', *Journal of Pharmacological and Experimental Therapeutics*, 124, 264–71.

MAIER, N.R.F. (1949) *Frustration: The Study of Behavior Without a Goal*, New York, McGraw-Hill.

MAIER, N.R.F., and GLASER, N.M. (1940) 'Studies of abnormal behavior in the rat:V. The inheritance of the "neurotic pattern" ', *Journal of Comparative Psychology*, 30, 413–18.

MASON, W.A. (1968) 'Early social deprivation in the non-human primates: implications for human behavior', in D.C. Glass (ed.), *Environmental Influences*, New York, Russell Sage Foundation, pp. 70–100.

MELZACK, R.A., and SCOTT, T.H. (1957) 'The effects of early experience on the response to pain', *Journal of Comparative and Physiological Psychology*, 50, 155–61.

MEYER-HOLZAPFEL, M. (1968) 'Abnormal behavior in zoo animals', in M.W. Fox (ed.), *Abnormal Behavior in Animals*, Philadelphia, Saunders, pp. 476–503.

MITCHELL, G. (1970) 'Abnormal behavior in primates', in L. Rosenblum (ed.), *Primate Behavior: Developments in Field and Laboratory Research*, New York, Academic Press.

MUCHA, R.F., and PINEL, J.P.J. (1979) 'Increased susceptibility to kindled seizures in rats following a single injection of alcohol', *Journal of Studies on Alcohol*, 40, 258–71.

NAQUET, R., and MELDRUM, B.S. (1972) 'Photogenic seizures in baboon,' in D.P. Purpura, J.K. Penry, D.B. Tower, D.M. Woodbury and R.D. Walter (eds), *Experimental Epilepsy. A Manual for the Laboratory Worker*, New York, Raven, pp. 373–406.

RANDRUP, A., and MUNKVAD, I. (1967) 'Stereotyped activities produced by amphetamine in several animal species and man', *Psychopharmacologia* (Berlin), 11, 300–310.

RANDRUP, A., and MUNKVAD, I. (1975) 'Stereotyped behavior', *Pharmacology and Therapeutics. Part B: General and Systemic Pharmacology*, 1, 757–68.

ROBBINS, T.W. (1982) 'Stereotypies: addictions or fragmented actions?', *Bulletin of the British Psychological Society*, 35, 297–300.

SCOTT, J.P. (1962) 'Critical periods in behavioral development', *Science*, 138, 949–58.

SILVERSTONE, T., and TURNER, P. (1982) *Drug Treatment in Psychiatry*, 3rd edition, London, Routledge & Kegan Paul.

SKINNER, B.F. (1948) ' "Superstition" in the pigeon', *Journal of Experimental Psychology*, 38, 168–72.

TEMERLIN, M.K. (1975) *Growing up Human*, Palo Alto, Science and Behavior Books.

THOMPSON, W.R., MELZACK, R., and SCOTT, T.H. (1956) ' "Whirling behavior" in dogs as related to early experience', *Science*, 123, 939.

Chapter 4 Behavioural anomalies and misdemeanours

AZRIN, N.H., HUTCHINSON, R.R., and HAKE, D.F. (1966) 'Extinction-induced aggression', *Journal of the Experimental Analysis of Behavior*, 9, 191–204.

BARNETT, S.A. (1955) ' "Displacement behavior" and "psychosomatic" disorder', *Lancet*, 2, 1203–8.

BLACKMAN, D.E. (1972) 'Conditioned anxiety and operant behaviour', in R.M. Gilbert and J.D. Keehn (eds), *Schedule Effects: Drugs, Drinking and Aggression*, Toronto, University of Toronto Press, pp. 26–49.

BOLLES, R.C. (1971) 'Species-specific defence reactions', in F.R. Brush (ed.), *Aversive Conditioning and Learning*, New York: Academic Press, pp. 183–233.

BRANCH, M.N., and DWORKIN, S.I. (1981) 'Effects of ratio contingencies on responding maintained by schedules of electric-shock presentation (response-produced shock)', *Journal of the Experimental Analysis of Behavior*, 36, 191–205.

BRELAND, K., and BRELAND, M. (1961) 'The misbehavior of organisms', *American Psychologist*, 61, 681–4.

BRELAND, K., and BRELAND, M. (1966) *Animal Behavior*, New York, Macmillan.

BRETT, L.P. and LEVINE, S. (1979) 'Schedule-induced polydipsia suppresses pituitary–adrenal activity in rats', *Journal of Comparative and Physiological Psychology*, 93, 946–56.

BROWN, J.S., MARTIN, R.C., and MORROW, M.W. (1964) 'Self-punitive behavior in the rat: facilitive effects of punishment on resistance to extinction', *Journal of Comparative and Physiological Psychology*, 57, 127–33.

BROWN, P.L., and JENKINS, H.M. (1968) 'Autoshaping of the pigeon's key-peck', *Journal of the Experimental Analysis of Behavior*, 10, 1–8.

BURKS, C.D., and FISHER, A.E. (1970) 'Anticholinergic blockade of schedule-induced polydipsia', *Physiology and Behavior*, 5, 635–40.

BYRD, L.D. (1969) 'Responding in the cat maintained by response-independent electric shock', *Journal of the Experimental Analysis of Behavior*, 12, 1–10.

BYRD, L.D. (1974) 'Modification of the effects of chlorpromazine and behavior in the chimpanzee', *Journal of Pharmacology and Experimental Therapeutics*, 189, 24–32.

COLOTLA, V.A. (1981) 'Adjunctive polydipsia as a model of alcoholism', *Neuroscience and Behavioral Reviews*, 5, 335–42.

COLOTLA, V.A., and BEATON, J.M. (1977) 'Effects of amphetamine on schedule induced drink durations in rats', *Mexican Journal of Behavior Analysis*, 3, 29–38.

DAVIS, H. (1979) 'Behavioral anomalies in aversive situations', in J.D. Keehn (ed.), *Psychopathology in Animals: Research and Clinical Implications*, New York, Academic Press, pp. 197–222.

DAVIS, H., and BURTON, J. (1974) 'The measurement of response force during a leverpress shock escape procedure in rats', *Journal of the Experimental Analysis of Behavior*, 22, 433–40.

EYSENCK, H.J. (1957) *The Dynamics of Anxiety and Hysteria*, London, Routledge & Kegan Paul.

FALK, J.L. (1961) 'Production of polydipsia in normal rats by an intermittent food schedule', *Science*, 133, 195–6.

FALK, J.L. (1964) 'Studies on schedule-induced polydipsia', in M.J. Wayner (ed.), *Thirst*, New York, Macmillan, pp. 95–116.

FALK, J.L. (1972) 'The nature and determinants of adjunctive behavior', in R.M. Gilbert and J.D. Keehn (eds), *Schedule Effects: Drugs, Drinking and Aggression*, Toronto, University of Toronto Press, pp. 148–73.

FISHER, A.E., and COURY, J.N. (1962) 'Cholinergic tracing of a central neural circuit underlying the thirst drive', *Science*, 138, 691–3.

FITZSIMONS, J.T. (1973) 'Some historical perspectives in the physiology of thirst', in A.N. Epstein, J.R. Kissileff and E. Stellar (eds), *The Neuropsychology of Thirst: New Findings and Advances in Concepts*, Washington, Winston, pp. 3–33.

FLORY, R.K. (1969) 'Attack behavior as a function of minimum inter-food interval', *Journal of the Experimental Analysis of Behavior*, 12, 825–8.

FREUD, S. (1936) *The Problem of Anxiety*, New York, Norton (trans. H.A. Bunker).

GILBERT, R.M. (1976) 'Drug abuse as excessive behaviour', *Canadian Psychological Review*, 17, 231-40.

GILLEN, N., and KEEHN, J.D. (1983) 'Effects of a conditioned suppression procedure on schedule-induced drinking by rats', *Psychological Record*, 33, 269–78.

GROSSMAN, S.P. (1962) 'Effects of adrenergic and cholinergic blocking agents on hypothalamic mechanisms', *American Journal of Physiology*, 202, 1230–6.

HARLOW, H.F. (1949) 'The formation of learning sets', *Psychological Review*, 56, 51–65.

HEARST, E., and JENKINS, H.M. (1974) *Sign-tracking: The Stimulus-Reinforcer Relation and Directed Behavior*, Austin, Psychonomic Society.

HINDE, R.A. (1962) 'Sensitive periods and the development of behaviour', in S.A. Barnett (ed.), *Lessons from Animal Behaviour for the Clinician*, London, Heineman.

HINDE, R.A. (1970) *Animal Behaviour: A Synthesis of Ethology and Comparative Psychology*, 2nd edition, New York, McGraw-Hill.

HINSON, R.E., and SIEGEL, S. (1982) 'Nonpharmacological bases of drug tolerance and dependence', *Journal of Psychosomatic Research*, 26, 495–503.

HOLZ, W.C., and AZRIN, N.H. (1961) 'Discriminative properties of punishment', *Journal of the Experimental Analysis of Behavior*, 4, 225–32.

HORNEY, K. (1937) *The Neurotic Personality of Our Time*, New York, Norton.

HOTHERSALL, D., and TUBER, D.S. (1979) 'Fears in companion dogs: characteristics and treatment', in J.D. Keehn (ed.), *Psychopathology in Animals: Research and Clinical Implications*, New York, Academic Press, pp. 239–55.

HUTCHINSON, R.R., and EMLEY, G.S. (1972) 'Schedule-independent factors contributing to schedule-induced phenomena', in R.M. Gilbert and J.D. Keehn (eds), *Schedule Effects: Drugs, Drinking and Aggression*, Toronto, University of Toronto Press, pp. 174–200.

JENKINS, H.M. (1975) 'Behavior theory to-day: a return to fundamentals', *Mexican Journal of Behavior Analysis*, 1, 39–54.

KEEHN, J.D., and COULSON, G.E. (1975) 'Schedule-induced choice of water versus alcohol', *Psychological Record*, 24, 325–8.

KEEHN, J.D., and MATSUNAGA, M. (1972) 'Attenuation of rats' alcohol consumption by trihexyphenidyl', in O. Forsander and K. Eriksson (eds), *Biological Aspects of Alcohol Consumption*, Helsinki, Finnish Foundation for Alcohol Studies.

KEEHN, J.D., and NAGAI, M. (1969) 'Attenuation of schedule-induced polydipsia by trihexyphenidyl', *Psychonomic Science*, 15, 61–2.

KEEHN, J.D., and RIUSECH, R. (1979) 'Schedule-induced drinking facilitates schedule-controlled feeding', *Animal Learning and Behavior*, 7, 41-4.

KEEHN, J.D., and STOYANOV, E. (1983) 'Disruption of adjunctive drinking by lick-dependent delays in feeding', *Psychological Record*, 33, 391–400.

KEEHN, J.D., COULSON, G.E., and KLIEB, J. (1976) 'Effects of haloperidol on schedule-induced polydipsia', *Journal of the Experimental Analysis of Behavior*, 24, 105–12.

LESTER, D., and FREED, E.X. (1972) 'A rat model of alcoholism?', *Annals of the New York Academy of Science*, 197, 54–9.

LEVIS, D.J. (1979) 'The infrahuman avoidance model of symptom maintenance and implosive therapy', in J.D. Keehn (ed.), *Psychopathology in Animals: Research and Clinical Implications*, New York, Academic Press, pp. 257–77.

LEVITT, R.A., and FISHER, A.E. (1966) 'Anticholinergic blockade of centrally induced thirst', *Science*, 154, 520–2.

LOGAN, F.A. (1972) 'Experimental psychology of animal learning and now', *American Psychologist*, 27, 1055–62.

McKEARNEY, J.W. (1968) 'Maintenance of responding under a fixed-interval schedule of electric shock presentation', *Science*, 160, 1249–51.

McKEARNEY, J.W. (1973) 'Effects of methamphetamine and chlordiazepoxide on schedule-controlled and adjunctive licking in the rat', *Psychopharmacologia*, 30, 375–84.

MILLENSON, J.R., and de VILLIERS, P.A. (1972) 'Motivational properties of conditioned anxiety', in R.M. Gilbert and J.R. Millenson (eds), *Reinforcement: Behavioral Analyses*, New York, Academic Press.

MORSE, W.H., MEAD, R.N., and KELLEHER, R.T. (1967) 'Modulation of elicited behavior by a fixed-interval schedule of electric shock presentation', *Science*, 157, 215–17.

MOWRER, O.H. (1947) 'On the dual nature of learning: a reinterpretation of "conditioning" and "problem solving" ', *Harvard Educational Review*, 17, 102–48.

PAVLOV, I.P. (1927) *Conditioned Reflexes*, Oxford, Oxford University Press.

PORSOLT, R.D., PICHON, M.L., and JALFRE, M. (1977) 'Depression. A new animal model sensitive to antidepressant treatments', *Nature*, 266, 730–2.

REYNOLDS, G.S., CATANIA, A.C. and SKINNER, B.F. (1963) 'Conditioned and unconditioned aggression in pigeons', *Journal of the Experimental Analysis of Behavior*, 6, 73–4.

SBORDONE, R., GARCIA, J., and CARDER, B. (1977) 'Shock-elicited aggression: its displacement by a passive social avoidance response', *Bulletin of the Psychonomic Society*, 9, 272–4.

SCHWARTZ, W., and WILLIAMS, D.R. (1972) 'Two different kinds of key peck in the pigeon: some properties of responses maintained by negative and positive response-reinforced contingencies', *Journal of the Experimental Analysis of Behavior*, 18, 201–16.

SEABROOK, M. (1982) 'The interaction between the personality and behaviour of the stockman and the behaviour and performance of dairy cows', paper presented at the 20th International Congress of Applied Psychology, Edinburgh.

SEGAL, E.F. (1972) 'Induction and the provenance of operants', in R.M. Gilbert and J.R. Millensen (eds), *Reinforcement: Behavioral Analyses*, New York, Academic Press.

SEGAL, E.F., ODEN, D.L. and DEADWYLER, S.A. (1965) 'Determinants of polydipsia: V. Effect of amphetamine and pentobarbital', *Psychonomic Science*, 3, 33–4.

SELIGMAN, M.E.P. (1975) *Helplessness*, San Francisco, Freeman.

SELIGMAN, M.E.P., and MAIER, S.F. (1967) 'Failure to escape traumatic shock', *Journal of Experimental Psychology*, 74, 1–9.

SIEGEL, R.K. (1979) 'Natural animal addictions: an ethological perspective', in J.D. Keehn (ed.), *Psychopathology in Animals: Research and Clinical Implications*, New York, Academic Press, pp. 29–60.

SKINNER, B.F. (1938) *The Behavior of Organisms*, New York, Appleton-Century-Crofts.

STEIN, L. (1963) 'Anticholinergic drugs and the central control of thirst', *Science*, 139, 46–8.

THOMPSON, T., BIGELOW, G., and PICKENS, R. (1971) 'Environmental variables influencing drug self-administration', in T. Thompson and R. Pickens (eds), *Stimulus Properties of Drugs*, New York, Appleton-Century-Crofts, pp. 193–207.

TINBERGEN, N. (1952) 'Derived activities: their causation, biological significance, origin and emancipation during evolution', *Quarterly Review of Biology*, 27, 1–32.

TUBER, D.S., HOTHERSALL, D., and VOITH, V.G. (1974) 'Animal clinical psychology: a modest proposal', *American Psychologist*, 29, 762–6.

ULRICH, R. (1967) 'Interaction between reflexive fighting and cooperative escape', *Journal of the Experimental Analysis of Behavior*, 10, 311–17.

WEISS, J.M., GLAZER, H.I., and POHORECKY, L.A. (1976) 'Coping behavior and neurochemical changes: an alternative explanation for the original "learned helplessness" experiments', in G. Serban and A. Kling (eds), *Animal Models in Human Psychobiology*, New York, Plenum.

YEROFEEVA (EROFEEW), M.N. (1916) 'Contribution to the study of destructive conditioned reflexes', *Comptes Rendus de la Societé Biologique*, 79, 239–40 (abstract).

Chapter 5 Experimental neurosis and psychophysiological disorders

ALEXANDER, F., FRENCH. J.M., and POLLOCK, G.H. (1968) *Psychosomatic specificity, Vol. 1. Experimental studies and results*, Chicago, University of Chicago Press.

BLANCHARD, E.B., and HERSEN, M. (1976) 'Behavioral treatment of hysterical neurosis: symptom substitution and symptom return reconsidered', *Psychiatry*, 39, 118–29.

BOUSFIELD, W.A. (1955) 'Lope de Vega and early conditioning', *American Psychologist*, 10, 828.

BOWDEN, D.M. (1976) Editor's introduction, in V.G. Startsev, *Primate Models of Human Neurogenic Disorders*, Hillsdale, N.J., Erlbaum, pp. 9–10.

BRADY, J., and HARRIS, A. (1977) 'The experimental production of obtained physiological states', in W.K. Honig and J.E.R. Staddon (eds), *Handbook of Operant Behavior*, Englewood Cliffs, Prentice-Hall, pp. 596–618.

BROADHURST, P.L. (1973) 'Animal studies bearing on abnormal behavior', in H.J. Eysenck (ed.), *Handbook of Abnormal Psychology*, 2nd edition, London, Pitman, pp. 721–54.

CROFT, P.G. (1951) 'Some observations of neurosis in farm animals', *Journal of Mental Science*, 97, 584–8.

DMITRUK, V.M. (1974) ' "Experimental neurosis" in cats: fact or artifact?', *Journal of Abnormal Psychology*, 83, 97–105.

ENGEL, B.T. (1981) 'Clinical biofeedback: a behavioral analysis', *Neuroscience and Behavioral Reviews*, 5, 397–400.

EROFEEW (YEROFEEVA), M.N. (1916) 'Contribution to the study of destructive conditioned reflexes', *Comptes Rendus de la Societé Biologique*, 79, 239–240 (abstract).

EYSENCK, H.J. (1957) *The Dynamics of Anxiety and Hysteria*, London, Routledge & Kegan Paul.

FOX, H.M. (1965) *Canine Behavior*, Springfield, Thomas.

GANTT, W.H. (1944) *Experimental Basis for Neurotic Behavior: Origin and Development of Artificially Produced Disturbances of Behavior in Dogs*, New York, Hoeber.

GANTT, W.H., and DYKMAN, R.A. (1957) 'Experimental psychogenic tachycardia', in P.H. Hoch and J. Zubin (eds), *Experimental Psychopathology*, New York, Grune & Stratton, pp. 12–19.

GERARD, M.W. (1948) 'The psychogenic tic in ego development', in F. Alexander and T.M. French (eds), *Studies in Psychosomatic Medicine: An Approach to the Cause and Treatment of Vegetative Disturbances*, New York, Ronald.

GRAY, J.A. (1964) *Pavlov's Typology: Recent Theoretical and Experimental Developments from the Laboratory of B.M. Teplov*, Oxford, Pergamon.

HANSEN, R.S. (1976) 'Nervousness and hysteria in mature female chickens', *Poultry Science*, 55, 531–43.

HEBB, D.O. (1947) 'Spontaneous neuroses in chimpanzees: theoretical relations with clinical and experimental phenomena', *Psychosomatic Medicine*, 9, 3–16.

HINDE, R.A. (1962) 'The relevance of animal studies to human neurotic disorders', in D. Richter, J.M. Tanner, L. Taylor and O.L. Zangwill (eds), *Aspects of Psychiatric Research*, New York, Oxford University Press, pp. 240–61.

HUNT, H.F. (1964) 'Problems in the interpretation of "experimental neurosis" ', *Psychological Reports*, 15, 27–35.

JONES, E. (1961) *The Life and Work of Sigmund Freud* (L. Trilling and S. Marcus, eds), New York, Basic Books.

KATZ, D. (1937) *Animals and Men: Studies in Comparative Psychology*, London, Longmans, Green.

KHANANASHVILI, M.M. (1976) 'Experimental neuroses in unrestrained animals', *Pavlovian Journal of Biological Science*, 11, 127.

KURTSIN, I.T. (1968) 'Pavlov's concept of experimental neurosis and abnormal behavior in animals', in M.W. Fox (ed.), *Abnormal Behavior in Animals*, Philadelphia, Saunders.

KURTSIN, I.T. (1976) *Theoretical Principles of Psychosomatic Medicine*, New York, Wiley.

LACHMAN, S.J. (1972) *Psychosomatic Disorders: A Behavioristic Interpretation*, New York, Wiley.

LAGUTINA, N.I., and SYSOEVA, A.F. (1969–70) 'Comparative physiological characteristics of experimental neuroses in vertebrates', *Soviet Psychology*, 8, 100-18.

LIDDELL, H.S. (1956) *Emotional Hazards in Animals and Man*, Springfield, Thomas.

MACMILLAN, M. (1963) 'Pavlov's typology', *Journal of Nervous and Mental Disease*, 137, 447–54.

MASSERMAN, J.H. (1943) *Behavior and Neurosis: An Experimental Psychoanalytic Approach to Psychobiologic Principles*, Chicago, University of Chicago Press.

MASSERMAN, J.H. (1950) 'Experimental Neuroses', *Scientific American*, 182, 38–43.

MASSERMAN, J.H. and SIEVER, P.W. (1944) 'Dominance, neurosis and aggression: an experimental study', *Psychosomatic Medicine*, 6, 7–16.

MERRILL, G.G. (1945) 'Breaking the killing habit in dogs by inhibiting the conditioned reflex', *Journal of the American Veterinary Medical Association*, 107, 69–70.

MIMINOSHVILI, D.I. (1960) 'Experimental neurosis in monkeys', in I.A. Usdin (ed.), *Theoretical and Practical Problems of Medicine and Biology in Experiments on Monkeys* (trans. R. Schacter), Oxford, Pergamon, pp. 53–67.

PARRY, H. (1948) 'Canine hysteria and wheat', *Lancet*, 1, 32–4.

PAVLOV, I.P. (1927) *Conditioned Reflexes*, Oxford, Oxford University Press.

PAVLOV, I.P. (1970) 'An attempt to understand the symptoms of hysteria physiologically', in W.H. Gantt, L. Pickenhain and Ch. Zwingmann (eds), *Pavlovian Approach to Psychopathology: History and Perspectives*, Oxford, Pergamon, pp. 20–34.

RAZRAN, G. (1971) *Mind in Evolution. An East–West Synthesis of Learned Behavior and Cognition*, Boston, Houghton-Mifflin.

REESE, W.G. (1978) 'Familial vulnerability for experimental neurosis', *Pavlovian Journal*, 13, 169–73.

ROZIN, P., and KALAT, J. (1962) 'Specific hungers and poison avoidance as adaptive specializations of learning', *Psychological Review*, 78, 459–86.

SANGER, V.L. and HAMDY, A.H. (1962) 'A strange fright–flight behavior pattern (hyşteria) in hens', *Journal of the American Veterinary Medical Association*, 140, 455–9.

SCOTT, J.P., and FULLER, J.L. (1965) *Genetics and the Social Behavior of the Dog*, Chicago, University of Chicago Press.

SELYE, H. (1956) *The Stress of Life*, New York, McGraw Hill.

SIDMAN, M. (1966) 'Avoidance behavior', in W.K. Honig (ed.), *Operant Behavior: Areas of Research and Application*, New York, Appleton, pp. 448–98.

SMART, R.G. (1965) 'Conflict and conditioned aversive stimuli in the development of experimental neuroses', *Canadian Journal of Psychology*, 19, 208–23.

SOLOMON, R.L., and WYNNE, L.C. (1954) 'Traumatic avoidance learning: the principles of anxiety conservation and partial irreversibility', *Psychological Review*, 61, 353–85.

SOLOMON, R.L., KAMIN, L.J., and WYNNE, L.C. (1953) 'Traumatic avoidance learning: the outcome of several extinction procedures with dogs', *Journal of Abnormal and Social Psychology*, 48, 291–302.

STARTSEV, V.G. (1976) *Primate Models of Human Neurogenic Disorders*, Hillsdale, N.J., Erlbaum, (trans. Marienne Schweinler and Vadim Pahn).

THOMAS, E., and DeWALD, L. (1977) 'Experimental neurosis: neurophysiological analysis', in J.D. Maser and M.E.P. Seligman (eds), *Psychopathology: Experimental Models*, San Francisco, Freeman, pp. 214–31.

WOLPE, J. (1952) 'Experimental neuroses as learned behaviour', *British Journal of Psychology*, 43, 243–68.

WOLPE, J. (1967) 'Parallels between animal and human neuroses', in J. Zubin and H.F. Hunt (eds), *Comparative Psychopathology: Animal and Human*, New York, Grune & Stratton, pp. 305–13.

WOODRUFF, R.A., GOODWIN, D.W., and GUZE, S.B. (1974) *Psychiatric Diagnosis*, New York, Oxford University Press.

WRIGHT, L. (1977) 'Conceptualizing and defining psychosomatic disorders', *American Psychologist*, 32, 625–8.

Chapter 6 Animals and addictions

ACTON, H.M.M. (1932) *The Last Medici*, London, Faber & Faber.

BOND, N.W., DiGIUSTO, E.L. (1977) 'Prenatal alcohol consumption and open-field behaviour in rats: effects of age at time of testing', *Psychopharmacology*, 52, 311–12.

CHEN, C-S. (1968) 'A study of the alcohol-tolerance effect and an introduction to a new behavioral technique', *Psychopharmacologia*, 12, 433–40.

CICERO, T.J. (1980) 'Animal models of alcoholism?' in K. Eriksson, J.D. Sinclair and K. Kiianmaa (eds), *Animal Models in Alcohol Research*, New York, Academic Press, pp. 99–117.

DAVIES, D.L. (1976) 'Definitional issues in alcoholism', in R.E. Tarter and A.A. Sugerman (eds), *Alcoholism. Interdisciplinary Approaches to an Enduring Problem*, Reading, Mass., Addison-Wesley, pp. 53–73.

ELLIS, F.W. and PICK, J.R. (1970) 'Experimentally-induced ethanol dependence in rhesus monkeys', *Journal of Pharmacology and Experimental Therapeutics*, 175, 88–93.

ELLISON, G.D. (1979) 'Animal models of psychopathology: studies in naturalistic colony environments', in J.D. Keehn (ed.), *Psychopathology in Animals: Research and Clinical Implications*, New York, Academic Press, pp.81–101.

ERIKSSON, K. (1980) 'Inherited metabolism and behavior towards alcohol: critical evaluation of human animal research', in K. Eriksson, J.D. Sinclair and K. Kiianmaa (eds), *Animal Models in Alcohol Research*, New York, Academic Press, pp. 3–20.

ERIKSSON, K., SINCLAIR, J.D., and KIIANMAA, K. (1980) *Animal Models in Alcohol Research*, New York, Academic Press.

ESSIG, C.F., and LAM, R.C. (1968) 'Convulsions and hallucinatory behavior following alcohol withdrawal in the dog', *Archives of Neurology*, 18, 626–32.

FALK, J.L., SAMSON, H.H., and WINGER, G. (1972) 'Behavioral maintenance of high concentrations of blood ethanol and physical dependence in the rat', *Science*, 177, 811–13.

FREEDMAN, D.X., and SENAY, E.C. (1973) 'Methadone treatment of heroin addiction', *Annual Review of Medicine*, 24, 153–164.

FREUND, G. (1969) 'Alcohol withdrawal syndrome in mice', *Archives of Neurology*, 21, 315–20.

FREUND, G. (1980) 'Comparison of alcohol dependence, withdrawal and hangover in humans and animals', in K. Eriksson, J.D. Sinclair and K. Kiianmaa (eds), *Animal Models in Alcohol Research*, New York, Academic Press, 293–308.

GOLDBERG, S.R., and SCHUSTER, C.R. (1967) 'Conditioned suppression by a stimulus associated with nalorphine in morphine-dependent monkeys', *Journal of the Experimental Analysis of Behavior*, 10, 235–42.

HINSON, R.E., and SIEGEL, S. (1980) 'The contribution of Pavlovian conditioning to ethanol tolerance and dependence', in H. Rigter and J.C. Crabbe (eds), *Alcohol Tolerance and Dependence*, Amsterdam, Elsevier, pp. 181–99.

ISBELL, H., FRASER, H.F., WIKLER, A., BELLEVILLE, R.E. and EISENMAN, A.J. (1955) 'An experimental study of the etiology of "rum fits" and delerium tremens', *Quarterly Journal of Studies on Alcohol*, 16, 1–33.

JONES, M.C. (1924) 'The elimination of children's fears', *Journal of Experimental Psychology*, 7, 383–90.

KALANT, H., LeBLANC, A.E., and GIBBINS, R.J. (1971) 'Tolerance to, and dependence on, some non-opiate psychotropic drugs', *Pharmacological Reviews*, 23, 135–91.

KEEHN, J.D. (1969) ' "Voluntary" consumption of alcohol by rats', *Quarterly Journal of Studies on Alcohol*, 30, 320–9.

KEEHN, J.D. and COULSON, G.E. (1975) 'Scheduled-induced choice of water versus alcohol', *Psychological Record*, 25, 325–8.

KEEHN, J.D., KUECHLER, H.A., OKI, G., COLLIER, D., and WALSH, R. (1973) 'Interpersonal behaviorism and community treatment of alcoholics', *Proceedings of the First Annual Alcoholism Conference of the National Institute of Alcohol Abuse and Alcoholism: Research on Alcoholism: Clinical Problems and Special Populations*, Rockville, Md, National Institute of Alcohol Abuse and Alcoholism, NIMH, pp. 153–76.

LESTER, D. (1961) 'Self-maintenance of intoxication in the rat', *Quarterly Journal of Studies on Alcohol*, 22, 223–231.

LESTER, D., and FREED, E.X. (1973) 'Criteria for an animal model of alcoholism', *Pharmacology, Biochemistry and Behavior*, 1, 103–7.

LEYHAUSAN, P. (1973) 'Addictive behavior in free ranging animals', in L. Goldberg and F. Hoffmeister (eds), *Psychic dependence*, Bayer-Symposium IV, New York, Springer-Verlag, pp. 58–64.

MAJCHROWICZ, E. (1975) 'Induction of physical dependence upon ethanol and the associated behavioral changes in rats', *Psychopharmacologia*, 43, 245–54.

MEISCH, R.A. (1980) 'Ethanol as a reinforcer for rats, monkeys and humans', in K. Eriksson, J.D. Sinclair and K. Kiianmaa (eds), *Animal Models in Alcohol Research*, New York, Academic Press, pp. 153–8.

MELLO, N.K. (1972) 'Behavioral studies of alcoholism', in B. Kissin and H. Begleiter (eds), *Biology of Alcoholism. Vol.2. Physiology and Behavior*, New York, Plenum.

MILLER, P.M. (1972) 'The use of behavioral contracting in the treatment of alcoholism: a case study', *Behavior Research and Therapy*, 3, 593–6.

MOTTIN, J.L. (1973) 'Drug-induced attenuation of alcohol consumption: review and evaluation of claimed, potential, or current therapies', *Quarterly Journal of Studies on Alcohol*, 34, 444–72.

O'BRIEN, C.P., TESTA, T., O'BRIEN, T., BRADY, J.P., and WELLS, B. (1977) 'Conditioned narcotic withdrawal in humans', *Science*, 195, 1000–2.

PAVLOV, I.P. (1927) *Conditioned Reflexes*, Oxford, Oxford University Press.

PIEPER, W.A., SKEEN, M.J., McCLURE, H.M., and BOURNE, P.G. (1972) 'The chimpanzee as an animal model for investigating alcoholism', *Science*, 176, 71–3.

RAY, O. (1978) *Drugs, Society and Human Behavior*, St. Louis, Mosby.

RODGERS, D.A. (1967) 'Alcohol preference in mice', in J. Zubin and H.F. Hunt (eds), *Comparative Psychopathology: Animal and Human*, New York, Grune & Stratton, pp. 154–201.

SAMSON, H.H., and FALK, J.L. (1974) 'Alteration of fluid preferences in ethanol-dependent animals', *Journal of Pharmacology and Experimental Therapeutics*, 190, 365–76.

SCHUSTER, C.R., and JOHANSON, C.E. (1974) 'The use of animal models for the study of drug abuse', in R.J. Gibbins (editor-in-chief), *Research Advances in Alcohol and Drug Problems. Vol.1*, New York, Wiley, pp.1–31.

SCHUSTER, C.R., RENAULT, P.F., and BLAINE, J. (1979) 'An analysis of the relationship of psychopathology to non-medical drug use', in R.W. Pickens and L.H. Heston (eds), *Psychiatric Factors in Drug Abuse*, New York, Grune & Stratton, pp. 1–19.

SIEGEL, R.K. (1979) 'Natural animal addictions: an ethological perspective', in J.D. Keehn (ed.), *Psychopathology in Animals: Research and Clinical Implications*, New York, Academic Press, pp. 29–60.

SIEGEL, R.K. and BRODIE, M. (1984) 'Alcohol self-administration by elephants', *Bulletin of the Psychonomic Society*, 22, 49–52.

SIEGEL, S. (1979) 'The role of conditioning in drug tolerance and addiction', in J.D. Keehn (ed.), *Psychopathology in Animals: Research and Clinical Implications*, New York, Academic Press, pp. 143–68.

SINCLAIR, J.D. (1974) 'Rats learning to work for alcohol', *Nature*, 249, 590–2.

STAINTON, H. (1941) 'Addiction in animals', *British Journal of Inebriety*, 41, 24–31.

TRICE, H.M. (1970) 'The alcoholic employee and supervisor: a general management problem', in R.E. Popham (ed.), *Alcohol and Alcoholism*, Toronto, University of Toronto Press, pp. 338–45.

WALLGREN, H., and BARRY, H. (1970) *Actions of Alcohol. Vol. 1. Biochemical, Physiological and Psychological Aspects*, Amsterdam, Elsevier.

WEEKS, J. (1962) 'Experimental morphine addiction: method for automatic intravenous injections in unrestrained rats', *Science*, 138, 143–4.

WOLPE, J. (1958) *Psychotherapy by Reciprocal Inhibition*, Stanford, Stanford University Press.

Chapter 7 True and model psychosis

AYLLON, T., and AZRIN, N.H. (1968) *The Token Economy: A Motivational System for Therapy and Rehabilitation*, New York, Appleton-Century-Crofts.

AYLLON, T., HAUGHTON, E., and HUGHES, H.B. (1965) 'Interpretation of symptoms: fact or fiction?', *Behaviour Research and Therapy*, 3, 1–7.

BERCEL, N.A. (1960) 'A study of the influence of schizophrenic serum on the behavior of the spider *Zilla-X-notata*', *Archives of General Psychiatry*, 2, 189–209.

BLEULER, E. (1950) *Dementia Praecox*, New York, International Universities Press.

CHERTOK, L. (1968) 'Animal hypnosis', in M.W. Fox (ed.), *Abnormal Behavior in Animals*, Philadelphia, Saunders.

COLOTLA, V.A. (1979) 'Experimental depression in animals', in J.D. Keehn (ed.), *Psychopathology in Animals: Research and Clinical Implications*, New York, Academic Press, pp. 223–38.

CONNELL, P.H. (1958) *Amphetamine Psychosis*, London, Chapman & Hall.

DAWKINS, M. (1980) *Animal Suffering: The Science of Animal Welfare*, London, Chapman & Hall.

ELLINWOOD, E.H., SUDILOVSKY, A., and NELSON, L. (1972) 'Behavioral analysis of chronic amphetamine intoxication', *Biological Psychiatry*, 4, 215–30.

ELLISON, G.D. (1977) 'Animal models of psychopathology: the low-norepinephrine and low-serotonin rat', *American Psychologist*, 32, 1036–45.

ELLISON, G.D. (1979) 'Animal models of psychopathology: studies in naturalistic colony environments', in J.D. Keehn (ed.), *Psychopathology in Animals: Research and Clinical Implications*, New York, Academic Press, pp. 81–101.

FERSTER, C.B. (1966) 'Animal behavior and mental illness', *Psychological Record*, 16, 345–56.

FERSTER, C.B. (1974) 'Behavioral approaches to depression', in R.J. Friedman and M.M. Katz (eds), *The Psychology of Depression: Contemporary Theory and Research*, Washington, D.C., Winston.

GALLUP, G.G. (1974) 'Animal hypnosis: factual status of a fictional concept', *Psychological Bulletin*, 81, 836–53.

GALLUP, G.G. and MASER, J.D. (1977) 'Tonic immobility: evolutionary underpinnings of human catalepsy and catatonia', in J.D. Maser and M.E.P. Seligman (eds), *Psychopathology: Experimental Models*, San Francisco, Freeman, pp. 334–57.

GROH, G., and LEMIEUX, M. (1968) 'The effect of LSD-25 on spider web formation', *International Journal of Addictions*, 3, 41–53.

HENN, F.A., and NASRALLAH, H.A. (1982) *Schizophrenia as a Brain Disease*, Oxford, Oxford University Press.

HOFFER, A., OSMOND, H., and SMYTHIES, J. (1954) 'Schizophrenia: a new approach II. Results of a year's research', *Journal of Mental Science*, 100, 29–45.

HOFMANN, A. (1959) 'Psychotomimetic drugs: chemical and pharmacological aspects', *Acta Physiologica Pharmacologica Neerlandica*, 8, 240–58.

JACOBS, B.L., TRULSON, M.E., and STERN, W.C. (1976) 'An animal behavior model for studying the actions of LSD and related hallucinogens', *Science*, 194, 741–3.

KETY, S.S. (1959) 'Biochemical theories of schizophrenia', *Science*, 129, 1528–32.

KUMAR, R. (1974) 'Animal models for evaluating psychotropic drugs', *Psychological Medicine*, 4, 353–9.

LYON, M., and ROBBINS, T.W. (1975) 'The action of central nervous system stimulant drugs: a general theory concerning amphetamine effects', in W. Essman and L. Valzelli (eds), *Current Developments in Psychopharmacology, Vol.2*, New York, Spectrum, pp. 89–163.

McKELLAR, P. (1957) *Imagination and Thinking*, New York, Basic Books.

MITCHELL, G. (1968) 'Persistent behavior pathology in rhesus monkeys following early social isolation', *Folia Primatologica*, 8, 132–47.

MITCHELL, J.M. (1953) 'A psychosis among cats', *Veterinary Record*, 65, 254.

NEWTON, J.E.O., and GANTT, W.H. (1968) 'History of a catatonic dog', *Conditional Reflex*, 3, 45–61.

OSMOND, H., and SMYTHIES, J. (1952) 'Schizophrenia: a new approach', *Journal of Mental Science*, 98, 309–15.

RANDRUP, A. and MUNKVAD, I. (1967) 'Stereotyped activities produced by amphetamine in several animal species and man', *Psychopharmacologia* (Berlin), 11, 300–10.

RANDRUP, A., and MUNKVAD, I. (1970) 'Biochemical, anatomical and psychological investigations of stereotyped behavior induced by amphetamines', in E. Costa and S. Garratini (eds), *Amphetamines and Related Compounds*, New York, Raven, pp. 695–713.

RANDRUP, A., and MUNKVAD, I. (1975) 'Stereotyped behavior', *Pharmacology and Therapeutics. Part B: General and Systemic Pharmacology*, 1, 757–68.

RATNER, C.S. (1967) 'Comparative aspects of hypnosis', in J.E. Gordon (ed.), *Handbook of Clinical and Experimental Hypnosis*, New York, Macmillan.

ROBBINS, T. (1976) 'Relationship between reward-enhancing and stereotypical effects of psychomotor stimulant drugs', *Nature*, 264, 57–9.

ROBBINS, T. and SAHAKIAN, B.J. (1980) 'Animal models of mania', in R. Belmaker and H. Van Praag (eds), *Mania: An Evolving Concept*, New York, Spectrum.

SELIGMAN, M.E.P. (1975) *Helplessness: On Depression, Development and Death*, San Francisco, Freeman.

SELIGMAN, M.E.P., and MAIER, S.F. (1967) 'Failure to escape traumatic shock', *Journal of Experimental Psychology*, 74, 1–9.

SILVERSTONE, T., and TURNER, P. (1982) *Drug Treatment in Psychiatry*, 3rd edition, London, Routledge & Kegan Paul.

SKINNER, B.F. (1957) 'The experimental analysis of behavior', *American Scientist*, 45, 343–71 (reprinted in *Cumulative Record*, New York, Appleton-Century-Crofts, 1961).

STEIN, L., and RAY, O.S. (1960) 'Brain stimulation reward "thresholds" self-determined in the rat', *Psychopharmacologia* (Berlin), 1, 251–6.

SUTHERLAND, N. (1976) *Breakdown*, London, Weidenfeld & Nicolson.

WEISS, J.M., GLAZER, H.I., and POHORECKY, L.A. (1976) 'Coping behavior and neurochemical changes: an alternative explanation for the original "learned helplessness" experiments', in G. Serban and A. Kling (eds), *Animal Models in Human Psychobiology, New York, Plenum.*

Chapter 8 Disorders of childhood

BAKWIN, H. (1942) 'Loneliness in infants', *American Journal of Diseases in Children*, 63, 30–40.

BAKWIN, H., and BAKWIN, R.M. (1966) *Clinical Management of Behavior Disorders in Children*, Philadelphia, Saunders.

BAREGGI, S.R., BECKER, R.E, GINSBURG, B., and GENOVESE, E. (1972) 'Paradoxical effect of amphetamine in an endogenous model of the hyperkinetic syndrome in a hybrid dog: correlation with amphetamine and p-hydroxyamphetamine blood levels', *Psychopharmacology*, 62, 217–24.

BERKSON, G. (1967) 'Abnormal stereotyped motor acts', in J. Zubin and H.F. Hunt (eds), *Comparative Psychopathology: Animal and Human*, New York, Grune & Stratton, pp. 76–94.

BOWLBY, J. (1969) *Attachment. Vol. 1*, New York, Basic Books.

BOWLBY, J. (1973) *Attachment and Loss. Vol. 2, Separation, Anxiety and Anger*, New York, Basic Books.

BOWLBY, J. (1976) 'Human personality development in an ethological light', in G. Serban and A. Kling (eds), *Animal Models in Human Psychobiology*, New York, Plenum, pp. 27–36.

BRADLEY, G. (1937) 'The behavior of children receiving benzedrine', *American Journal of Psychiatry*, 94, 577–85.

CAMPBELL, B.A., and RANDALL, P.K. (1975) 'Paradoxical effects of amphetamine on behavioral arousal in neonatal and adult rats: a possible animal model of the calming effect of amphetamine on hyperkinetic children', in N.R. Ellis (ed.), *Aberrant Development in Infancy: Human and Animal Studies*, Hillsdale, N.J., Erlbaum, pp. 105–12.

COLE, S.O., and MOORE, S.F. (1976) 'The hyperkinetic child syndrome: the need for reassessment', *Child Psychiatry and Human Development*, 7, 103-12.

CONNERS, C.K., EISENBERG, L., and BARCAI, A. (1967) 'Effects of dextroamphetamine on children', *Archives of General Psychiatry*, 17, 478–85.

CORSON, S.A., CORSON, E.O., ARNOLD, L.E., and KNOPP, W. (1976) 'Animal models of violence and hyperkinesis. Interaction of psychopharmacologic and psychosocial therapy in behavior modification', in G. Serban and A. Kling (eds), *Animal Models in Human Psychobiology*, New York, Plenum, pp. 111–39.

CROSS, H.A., and HARLOW, H.F. (1965) 'Prolonged and progressive effects of partial isolation on the behavior of macaque monkeys', *Journal of Experimental Research on Personality*, 1, 39–49.

DAVENPORT, R.K. and MENZEL, E.W. (1963) 'Stereotyped behavior of the infant chimpanzee', *Archives of General Psychiatry*, 8, 99–104.

DENNIS, W. (1973) *Children of the Creche*, New York, Appleton-Century-Crofts.

ENGEL, G.L., and REICHSMAN, F. (1956) 'Spontaneous and experimentally induced depressions in an infant with a gastric fistula. A contribution to the problem of depression', *Journal of the American Psychoanalytic Association*, 4, 428–52.

FULLER, J.L. (1953) 'Cross-sectional and longitudinal studies of adjustive behavior in dogs', *Annals of the New York Academy of Sciences*, 56, 214–24.

GOLDBERG, A.M., and SILBERGELD, E.K. (1977) 'Animal models of hyperactivity', in. I. Hanin and E. Usdin (eds), *Animal Models in Psychiatry and Neurology*, Oxford, Pergamon, pp. 371–84.

HARLOW, H.F. (1958) 'The nature of love', *American Psychologist*, 13, 673–85.

HARLOW, H.F., and HARLOW, M.K. (1971) 'Psychopathology in monkeys', in H.D. Kimmel (ed.), *Experimental Psychopathology: Recent Research and Theory*, New York, Academic Press, pp. 203–29.

HINDE, R.A. (1972) *Social Behaviour and its Development in the Sub-human Primates*, Eugene, Oregon, Oregon State System of Higher Education.

HINDE, R.A. (1976) 'Workshop I', in G. Serban and A. Kling (eds), *Animal Models in Human Psychobiology*, New York, Plenum, p. 64.

HINDE, R.A., and SPENCER-BOOTH, Y. (1970) 'Individual differences in the responses of rhesus monkeys to a period of separation from their mothers', *Journal of Child Psychology and Psychiatry*, 11, 159–76.

KAUFMAN, I.G. (1973) 'Mother–infant separation in monkeys: an experimental model', in J.P. Scott and E.C. Senay (eds), *Separation and Depression: Clinical and Research Aspects*, Washington, American Association for the Advancement of Science, pp. 33–52.

KAUFMAN, I.G., and ROSENBLUM, L.A. (1967a) 'Depression in infant monkeys separated from their mothers', *Science*, 155, 1030–1.

KAUFMAN, I.G., and ROSENBLUM, L.A. (1967b) 'The reactions to separation in infant monkeys: anaclitic depression and conservation-withdrawal', *Psychosomatic Medicine*, 29, 648–75.

KRASNOGORSKI, N.I. (1925) 'The conditioned reflexes and children's neuroses', *American Journal of Disorders of Children*, 30, 754.

KRUSHINSKII, L.V. (1962) *Animal Behavior: Its Normal and Abnormal Development*, New York, Consultants' Bureau.

LEVISON, C.A. (1970) 'The development of headbanging in a young rhesus monkey', *American Journal of Mental Deficiency*, 74, 323–8.

LIPSITT, L.P. (1979) 'Critical conditions in infancy: a psychological perspective', *American Psychologist*, 34, 973–80.

LYON, M., and NIELSON, E.B. (1979) 'Psychosis and drug-induced stereotypies', in J.D. Keehn (ed.), *Psychopathology in Animals: Research and Clinical Implications*, New York, Academic Press, pp. 103–42.

LYON, M., and ROBBINS, T.W. (1975) 'The action of central nervous system stimulant drugs: a general theory concerning amphetamine effects', in W. Essman and L. Valzelli (eds), *Current Developments in Psychopharmacology, Vol. 2*, New York, Spectrum, pp. 89–163.

MASON, W.A. (1968) 'Early social deprivation in the non-human primates: implications for human behavior', in D.C. Glass (ed.), *Environmental Influences*, New York, Russell Sage Foundation, pp. 70–100.

MASON, W.A. (1970) 'Motivational factors in psychosocial development', in W.J. Arnold and M.M. Page (eds), *Nebraska Symposium on Motivation*, Lincoln, University of Nebraska Press, pp. 35–67.

MICHAELSON, I.A., BORNSCHEIN, R.L., LOCH, R.K., and RAFALES, L.S. (1977) 'Minimal brain dysfunction hyperkinesis: significance of nutritional states in animal models of hyperactivity', in I. Hanin and E. Usdin (eds), *Animal Models in Psychiatry and Neurology*, Oxford, Pergamon, pp. 37–49.

POLAN, C.G., and SPENCER, B.L. (1959) 'A check list of symptoms for autism of early life', *West Virginia Medical Journal*, 55, 198–204.

SCOTT, J.P., and FULLER, J.L. (1965) *Genetics and the Social Behavior of the Dog*, Chicago, University of Chicago Press.

SCOTT, J.P., STEWART, J.M., and DeGHETT, V.J. (1973) 'Separation in infant dogs: emotional response and motivational consequences', in J.P. Scott and E.C. Senay (eds), *Separation and Depression: Clinical and Research Aspects*, Washington, American Association for the Advancement of Science, pp. 3–32.

SECHZER, J.A. (1977) 'The neonatal split-brain kitten: a laboratory analogue of minimal brain dysfunction', in J.D. Maser and M.E.P. Seligman (eds), *Psychopathology: Experimental Models*, San Francisco, Freeman, pp. 308–33.

SPITZ, R.A. (1945) 'Hospitalism. An inquiry into the genesis of psychiatric conditions in early childhood', *Psychoanalytic Study of the Child*, 1, 53–74.

SPITZ, R.A., and WOLF, K.M. (1946) 'Anaclitic depression: an inquiry into the genesis of psychiatric conditions in early childhood, II', *Psychoanalytic Study of the Child*, 2, 313–42.

STEPHENS, D.B. (1974) 'Studies of the effect of social environment on the behaviour and growth rates of artificially removed British Friesian male calves', *Animal Production*, 18, 23–34.

TEMERLIN, M.K. (1975) *Growing Up Human*, Palo Alto, Science and Behavior Books.

TINBERGEN, E.A., and TINBERGEN, N. (1972) 'Early childhood autism: an ethological approach', *Journal of Comparative Ethology*, 10 (supplement).

TINBERGEN, N. (1953) *The Herring Gull's World*, London, Collins.

WATSON, J.B. (1924) *Behaviorism*, Chicago, University of Chicago Press.

WATSON, J.B., and RAYNER, R. (1920) 'Conditioned emotional reactions', *Journal of Experimental Psychology*, 3, 1–14.

WENDER, P.H. (1971) *Minimal Brain Dysfunction in Children*, New York, Wiley.

YULE, W. (1978) 'Diagnosis: developmental psychological assessment', in A.F. Kalverboer, H.M. vanPraag and J. Mendlewicz (eds), *Minimal Brain Dysfunction: Fact or Fiction. Advances in Biological Psychiatry, Vol. 1*, Basel, Karger, pp. 35–54.

Chapter 9 Summary and conclusions

DOMINO, E.F. (1976) 'Indole hallucinogens as animal models of schizophrenia', in G. Serban and A. Kling (eds), *Animal Models in Human Psychobiology*, New York, Plenum, pp. 239–59.

ELLISON, G.D. (1979) 'Animal models in psychopathology: studies in naturalistic environments' in J.D. Keehn (ed.), *Psychopathology in Animals: Research and Clinical Implications*, New York, Academic Press, pp. 81–101.

HINDE, R.A. (1962) 'The relevance of animal studies to human neurotic disorders', in D. Richter, J.M. Tanner, L. Taylor and O.L. Zangwill (eds), *Aspects of Psychiatric Research*, New York, Oxford University Press, pp. 240–61.

Subject index

Name index